3rd Annual World Congress on Sustainable Engineering 2013

Topical Conference at the 2013 AIChE Annual Meeting: Global Challenges for Engineering a Sustainable Future

San Francisco, California, USA
3-8 November 2013

ISBN: 978-1-63439-028-6

Printed from e-media with permission by:

Curran Associates, Inc.
57 Morehouse Lane
Red Hook, NY 12571

Some format issues inherent in the e-media version may also appear in this print version.

Printed by Curran Associates, Inc. (2014)

For permission requests, please contact AIChE
at the address below.

AIChE
3 Park Avenue
New York, NY 10016-5991

Phone: (203) 702-7660
Fax: (203) 775-5177

www.aiche.org

Additional copies of this publication are available from:

Curran Associates, Inc.
57 Morehouse Lane
Red Hook, NY 12571 USA
Phone: 845-758-0400
Fax: 845-758-2634
Email: curran@proceedings.com
Web: www.proceedings.com

TABLE OF CONTENTS

(32a) Overview Of Materials In The Built Environment

Judkoff, R.,

Originally presented on: 11/4/2013 8:30:00 - 8:52:00

Materials advances could help to reduce the energy and environmental impacts of buildings. Globally, buildings consume 30%-40% of primary energy and account for 25%-33% of CO_2 emissions. This talk explores some foundational concepts about how energy is used in buildings, and also highlights areas where materials advances would be most beneficial from an energy perspective.

(32b) Energyplus Model-Based Predictive Control (EPMPC) Using Matlab/Simulink and Mle+

Zhao, J., Center for Building Performance and Diagnostics
Lam, K. P., Center for Building Performance and Diagnostics
Ydstie, B. E., Carnegie Mellon University
Originally presented on: 11/4/2013 8:52:00 - 9:14:00

This paper proposes the EnergyPlus Model-based Predictive Control (EPMPC) system to control under floor air distribution (UFAD) in a multi-zone two stories open office building – The Center for Sustainable Landscapes – at Phipps Conservatory in the city of Pittsburgh. The building is equipped with a central air-handling unit with ground source heat pump as the cooling and heating source. The UFAD is used for the open office, conference rooms and other occupied spaces. Ceiling-based air distribution system is used for service spaces, such as restrooms, mechanical rooms, and storage rooms. The ASHRAE 90.1-2007 baseline and design phase whole building models are created using DesignBuilder and EnergyPlus program under the "Design-Build-Operate Energy Information Modeling (DBO-EIM)" infrastructure. The key of the DBO-EIM infrastructure is to use the detailed design phase energy model throughout the entire building life cycle by validating and adapting model assumptions during construction, commissioning and operation periods. Therefore, the model is expected to be "accurate" and easy-to-use for model-based building control systems in the daily operation to optimize energy and comfort. In this paper, the design phase model is applied in both the baseline and the EPMPC control. The EPMC controller includes both continuous and discrete variables and it is not possible to use optimization algorithms that rely on gradient search. In order to solve the optimization problem we developed an exhaustive search optimization algorithm within the Matlab/Simulink environment via maximum likelihood estimation+, co-simulation tool. In two weeks of winter and summer simulation using *typical meteorological year 3* weather file, comparing to the baseline logic control, which is also implemented using the same co-simulation method, the EPMPC can save HVAC operation energy by 18.9%. The occupant *predicted mean vote* value is as good as the baseline if not better. The benefits of using the EnergyPlus model over the simplified first principle or identified linear model include: (1) it can model complex buildings and systems; (2) it can be continuously updated with little cost and effort; (3) it requires less mathematics skills to manage in daily operation once the system has been developed.

(32c) The Sol-Char Toilet: Concentrated Solar Power Transported By Fiber Optics for Pyrolysis of Human Waste to Biochar

Fisher, R. P., University of Colorado at Boulder
Lewandowski, A., National Renewable Energy Lab
Yacob, T. W., University of Colorado at Boulder
Mahoney, R. B., University of Colorado at Boulder
Mejic, D., University of Colorado at Boulder
Oversby, C., University of Colorado at Boulder
Beck, S., University of Colorado at Boulder
Ward, B. J., University of Colorado at Boulder
Kearns, J., University of Colorado at Boulder
Summers, R. S., University of Colorado
Linden, K. G., University of Colorado at Boulder
Weimer, A. W., University of Colorado at Boulder
Originally presented on: 11/4/2013 9:14:00 - 9:36:00

Currently, 2.6 billion people on the planet do not have access to adequate sanitation and many communities in the developing world have neither the facilities nor the resources to properly treat human waste, leading to open defecation posing enormous public health risks as well as issues related to individual dignity and safety. Sanitation is not a problem unique to developing countries – modern sanitation processes are also unsustainable due to their often massive requirements for energy and water. The Sol-Char Toilet, a solar waste treatment technology developed at the University of Colorado Boulder, takes advantage of our most abundant renewable resource: the sun. In the Sol-Char Toilet, concentrated solar power (CSP) is transmitted through fiber optic cables to drive pyrolysis of human waste, converting a hazardous and unusable material to biochar. Biochar is a carbonized solid product of thermal biomass decomposition consisting largely of condensed aromatic zones that when applied as a soil amendment imparts agronomic benefits and is recalcitrant over a long timescale (10^2 – 10^3years). Biochar presents a solution relevant to the world today: reducing greenhouse gas emissions by sequestering carbon in biochar for several hundreds of years or even millennia. The over-reaching goal of the project is to develop a cost-effective, self-contained, and sustainable sanitation process by which human waste can be converted to useful products without connection to water, sewer or electricity.

Sunlight is concentrated to nearly 2000 Suns by a parabolic mirror, directed into a fiber optic bundle, and routed to a solar hood that directly irradiates a pyrolysis reactor. Two identical reactor vessels are switched between a pyrolysis location under the solar hood and a toilet location via a rotating carousel mechanism, thus achieving a semi-continuous pyrolysis process during daylight hours. The Sol-Char Toilet prototype represents a novel application of CSP to the areas of environmental, sanitation, and reaction engineering. This presentation will focus on the design and operation of the 1 kW Sol-Char Toilet prototype designed and constructed at the University of Colorado at Boulder.

(32d) Building-Integrated Active Modular Phytoremediation System

Dyson, A., Rensselaer Polytechnic Institute/Center for Architecture Science and Ecology
Aydogan Akseli, A., Rensselaer Polytechnic Institute
Vollen, J., Rensselaer Polytechnic Institute
Nyman, M., Rensselaer Polytechnic Institute
Bird, J.,
Originally presented on: 11/4/2013 9:36:00 - 9:58:00

Contemporary construction materials and mechanical building ventilation systems types are complicit in the degradation of indoor air quality (IAQ) and have become principal contributors to public health concerns in developed countries. These challenges are further exacerbated by the increasing cost of energy that is required to maintain healthy levels of IAQ in these buildings. Addressing these issues will require not only innovative technologies and approaches, but also deep collaborations across several disciplines.

This research stems from an interdisciplinary framework for the development of a building-integrated phytoremediation system using bio-mechanical techniques to 'scrub' toxins from the air and maintain desired psychometric profiles for IAQ by integrating modules containing hydroponic-supported plant materials into heating, ventilating and air conditioning (HVAC) systems. This research is multi-dimensional and interdisciplinary in scope, with the precise goal of integrating lab-based testing protocols to reiterative architectural systems design parameters.

There are three scopes that will be presented within this research: 1.Parametric framework for information exchange across disciplines; 2.Acqusition of scientific data on the sorption of VOCs (*e.g.,* formaldehyde and benzene) to growing media and granular activated carbon, removal efficiency of the selected VOCs from air utilizing the building-integrated phytoremediation system and the effect of phytoremediation as a bio-mechanical technique to remove formaldehyde and benzene from indoor air and; 3. Co-design of an integrated bio-mechanical HVAC system at building-scale. A series of experiments were undertaken to determine the performance parameters for architectural integration, including: the toxin removal capacity of various growing media and plant species; the impact of module morphology on contaminant concentration levels and air flow rates; and the rates of remediation by constituent components within plant leaves and the root rhizosphere. The conclusions from these experiments show that the building-integrated active modular phytoremediation system (AMPS) could have a significant impact in contributing to remediation strategies addressing poor IAQ in buildings and energy reduction through improved HVAC performance.

(32e) Integrating Technological and Ecological Design for Sustainable Buildings On a University Campus

Gibbemeyer, E. L., The Ohio State University
Bakshi, B. R., The Ohio State University
Originally presented on: 11/4/2013 9:58:00 - 10:20:00

Importance

The Ohio State University has pledged to become carbon neutral by 2050, an ambitious and currently out of reach goal. Technological systems often find the task of carbon neutrality to be challenging and frequently impossible. Therefore, we look to nature and ecosystems as examples of systems that have maintained sustainable levels of carbon, as well any other resource we care to examine. Thus, instead of merely optimizing a technological system to have minimum carbon emissions, we also consider the surrounding ecosystem and optimize the entire system to minimize carbon in what we call "techno-ecological networks." For example, in optimizing a building, our technological system, we can optimize it to be as close to carbon neutrality as possible; however, by simply planting additional trees outside of it, we can further reduce its carbon footprint by accounting for the carbon these trees will sequester. Including the ecosystem in our network allows us to reach a smaller carbon footprint that was previously infeasible when only looking at the technological components of our system.

Originality

While there have been many studies evaluating and optimizing the Ecological Footprints of buildings, ecosystems are usually not included in the evaluation. This means that any optimized solutions found in these studies could be further improved by also including the surrounding ecosystems. We also plan to include behavioral variables such as choosing a thermostat set point. Previously, we have incorporated these ideas into a simplified residential system; in this project, our techno-ecological network will be the real system of a segment of the OSU Columbus Campus that will include buildings and their surrounding green space. This real system will allow us to continue testing the effects of including ecological and behavioral variables into this optimization framework of minimizing carbon footprint, water footprint, and cost.

Methods Employed

We will use software developed by the DOE to model our campus buildings, and the models will be verified using data collected by the OSU Energy Services and Sustainability office. We then will begin making changes to technological variables to see what retrofitting options would lower our carbon and water footprints, as well as adding in the ecological variables through biophysical models including InVEST and CENTURY. Once we have developed all the different variables, we will use optimization software to solve for scenarios that minimize carbon, water, and cost.

Results

Because our optimization is multi-objective, we will not simply obtain a single solution; we will find a Pareto optimal set of solutions. However, we anticipate some variables to lead to "win-win" scenarios, such as installing a rainwater capture system to lower water footprint and also lower water utility costs. In finding "win-win" scenarios, we hope to find opportunities for changes that can be made and allow this study to turn into a true "living laboratory" where students can become more ecologically conscious. Nonetheless, for other variables, we will require a judgment call on the relative importance of our objectives if we want to be able to select a "best" scenario that we could implement on campus.

(32f) Energy Conservation Approach to Cooling Process By Employing Preheat Tank and Chilled Water Tanks

Peters, D. R. W., University of Alabama at Birmingham
Li, Z., University of Alabama at Birmingham
Originally presented on: 11/4/2013 10:20:00 - 10:42:00

Energy conservation has been a hot topic worldwide. With the rapid growth of population and urbanization, the energy demand is increasing annually. According to the 2005 *Annual Energy Review* compiled by the Energy Information Administration (EIA) within the U.S. Department of Energy (DOE), the energy consumption in the United States including transportation, residential, and commercial end uses, shows a trend of increasing during the period of 1970 to 2005. Over the years from 1970 to 2004, total commercial end uses have an annual growth rate of 2.6 percent, and electricity consumption accounts for 76.2% of total commercial energy consumption. The HVAC system is an important process to provide comfortable temperatures, but demanding a large amount of energy and mainly contributing to the peak electricity consumption. The chilled water tank is a technology that can dramatically reduce the peak energy consumption with installation in traditional air conditioning systems, especially for the hot and humid climates. The preheat tank is a device used for heat recovery purposes that absorbs the heat from air flows to decrease the air temperature entering the chilled water tank and enhance the performance of cooling with even less energy requirements.

Our research utilizes the preheat tank and chilled water tank in commercial and conventional air conditioning systems to achieve energy conservation opportunities in Birmingham, Alabama. The preheat tank was installed to exchange heat with hot air, and the heated water can be used for lavatory and other purposes. The chilled water tank was set up to provide cooling with less energy consumption. The Generalized Fluid System Simulation Program (GFSSP) software was utilized to analyze the air conditioning system involving the heat recovery tank and chilled water tank. In addition, experiments were conducted in a small building at the University of Alabama at Birmingham campus with an installed small preheat tank and chilled water tank, respectively, to provide data needed for software programming. The energy consumption of the modified HVAC system was monitored to check the energy conservation rate. Some factors that may affect the performance of the modified system, include ambient temperature, humidity, water temperature in the tank, air conditioning set temperature, and dew point. The variables were recorded and analyzed in regression models to determine which factor(s) are significant for energy conservation. Moreover, covariance analysis was performed to determine the relationship between these factors. The results are expected to provide energy conservation for air conditioning systems, especially during the peak hours of energy consumption in hot and humid summer climates which is the typical weather condition in Birmingham, Alabama. Finally, the payback period was estimated by comparing energy bill savings and capital investment. The research can be extended to other southern states that experience hot and humid climates, such as Texas, Arizona, Georgia, and Florida.

Environmental Division

(32g) Operation of Residential Hybrid Renewable Energy Systems: Integrating Forecasting, Optimization and Demand Response

Wang, X.,
El-Farra, N. H., University of California, Davis
Palazoglu, A., UC Davis

Originally presented on: 11/4/2013 10:42:00 - 11:04:00

Hybrid renewable energy systems, which consist of solar, wind and other energy generation and storage units, have been widely researched in recent years owing to the reduction in traditional fossil resources and increase in the world-wide energy demand. This paper focuses on the optimal design and operation of a standalone community-scale hybrid renewable energy system comprising photovoltaic (PV) cells, wind turbines, diesel generators and storage units such as batteries. A detailed modeling of each system component is presented. The demand-side management as well as the performance of the generation system are considered to guarantee that the electricity demand of the consumer is satisfied and the overall system economic and environmental cost is minimized.

To optimally allocate electricity generation among different units, a precise prediction of the future electricity consumption and the available solar and wind resources is necessary. Both the day-ahead and real-time predictions are made to provide sufficient information for further optimization. For the day-ahead prediction, historical demand as well as solar radiation and wind speed data are used as the basis for the classification of day types, which are determined by the temperature, humidity, seasonality and other related characteristics. During the day-ahead operational management, a general electricity usage and a generation profile are developed to realize the preliminary dispatch strategy. For the real-time prediction, autoregressive and polynomial fitting methods are adopted to facilitate time-series analysis and forecasting. The online forecasting model is able to consider the updated data and modify the overall prediction and optimization.

On the demand side, a demand-responsive scheme is integrated with the operation of power generation. Among all the residential electricity usage, some are rigid in demand which cannot be changed or shifted such as lighting and cooking while others are flexible loads including washing machines, clothes dryers and vacuum cleaners, which are the key factors to be considered to manage the demand side. These loads can be arranged to reshape demand profiles to match renewable output better.

In the optimization stage, the method of model predictive control (MPC) is adopted by considering real-time optimization to minimize multiple objectives over a receding horizon, the most significant of which is to minimize the generation cost within the time window considered. The environmental influences are also accounted for by the indicators of equivalent cost. As constraints, each energy resource has its maximum generation capacity, as well as minimum output level, which decides the operation range of the generation unit. Limits of ramping rate of each resource are also posed to avoid damage caused by excessively rapid change of power output. The formulation is simplified to a convex optimization problem. At each time interval, only the first set of power references calculated by the optimization are implemented. The optimal allocation between different resources drives the generation to meet electricity requirement of the demand side, with minimum operating cost and environmental impact. Different strategies to realize the receding-horizon method are applied and the results are compared. It turns out that the length of the moving horizon has a critical impact on the optimal allocation of electricity generation among the available resources. Proper selection of the horizon allows for an efficient and optimal operation of the energy system.

The strength of the proposed strategy is that it is able to achieve global optimization over a certain time period and at the same time the users are able to obtain sufficient information of the future conditions and thus make rational decisions about electricity management. Compared with traditional operation of energy systems, the on-line process that combines real-time prediction, optimization and demand-responsive scheme will improve system efficiency and take better advantage of all available resources.

We will show several simulation and optimization case studies that demonstrate the effectiveness and applicability of the proposed receding horizon scheme. Future work to connect the current system to the electrical smart-grid to realize large-scale application will also be discussed.

(36a) Green Chemistry, Green Engineering, Green Economy At US EPA 's Sustainable Technologies Division

Leazer, J., U.S. Environmental Protection Agency
Originally presented on: 11/4/2013 8:35:00 - 8:50:00

Green Chemistry and Green Engineering Research in the sustainable Technologies Division of USEPA's National Risk Management Research Laboratory is focused on reducing the adverse human health and environmental impacts of the synthesis of economically important chemicals and chemical processes. Our research focuses on several categories including:

1) Sustainable Materials Management, which accounts for ecological scarcity and resource efficiency;

2) Green Chemistry, which focuses on the minimization/elimination of human health and environmental impacts as well as resource efficiency;

3) Green Engineering, which focuses on energy and resource conservation;

4) Sustainability Measurement, which allows us to determine whether a system is becoming more sustainable or not;

5) Life Cycle Assessment, ensuring minimal impact across the entire life cycle of a product/good/service and

6) Supply Chain Design, which looks at the optimization and increased sustainability of the entire supply chain of a product.

These research areas have the common goal of green production of economically viable chemicals and chemical processes.

(36b) Sustainability Practice in Industry

Powell, J. B., Shell Westhollow Technology Center

Originally presented on: 11/4/2013 8:50:00 - 9:05:00

This seminar will address how sustainability is implemented in industry, with real world examples from the energy and chemical processing industries. Topics will include making the case for sustainability, sustainability metrics and reporting, the importance of considering sustainability early during new process or product development, and how industry may be organized to incorporate sustainability principles into decision making. The importance of external stakeholder engagement in project planning and execution will be addressed.

In industry, sustainability is viewed as a key enabler for license to operate, and can be seen as a driver for much of the new innovation and development in our profession. Competency in life cycle inventory and overall environmental footprint and social impact assessment has become increasingly important in assessing preferred opportunities for technology development and business deployment. Demand for excellence in safe and responsible implementation and performance will continue strong for the developed world, and dissemination of best technologies and practices to developing economies will be critical for addressing global sustainability needs and environmental challenges.

(36c) Green Engineering and Sustainability in the Chemical Engineering Curriculum

Shonnard, D. R., Michigan Technological University
Allen, D. T., The University of Texas at Austin
Nguyen, N., U.S. Environmental Protection Agency
Austin, S., Chemical Engineering Branch, US EPA
Originally presented on: 11/4/2013 9:05:00 - 9:20:00

A green economy will be well-served by an educated chemical engineering workforce firmly grounded in green and sustainable engineering concepts. The goal of this presentation is to outline updated green engineering and sustainable engineering content for undergraduate and graduate chemical engineering education. The work of chemical engineers is central to the development of sustainable chemical and energy production systems and to meet global human needs. Incorporating environmental issues and sustainability into the design of engineered products, processes, and systems generally means evaluating environmental, economic, and social performance. This presentation will summarize two approaches to incorporate green and sustainable engineering into undergraduate chemical engineering education; **i.** a stand-alone elective course, and **ii.** modules for required core chemical engineering courses. A detailed overview of chapter-by-chapter content will also be presented with tie-ins to core courses.

(36d) Driving Private Sector Adoption of Greener Processes and Technologies

Lindsay, A., Author: Abby Lindsay, American University, Doctoral Candidate

Originally presented on: 11/4/2013 9:40:00 - 9:55:00

In addition to advanced environmental technologies and improved chemical processes, a green economy will require widespread adoption by the private sector. This presentation will outline ways in which greener practices are spread to a wider number of companies through focusing on experiences throughout Latin America. Most larger companies with extensive resources and access to financing have realized the environmental and cost benefits to reducing energy and water use, decreasing pollution, and increasing water and materials efficiency. Micro, small, and medium-sized enterprises (MSMEs) can benefit just as much, but often lack the awareness, financial resources, or skills to increase their efficiency. Given that the majority of companies are MSMEs, reducing the resource-use and pollution intensity of these companies is critical to a green economy. This presentation will look at factors that drive the diffusion of environmental processes and technologies, and determinants required for successful adoption, such as adequate engineering education systems, availability of financing, and policy incentives.

Environmental Division

(36e) Minimization of the GHG Emissions At the Macroeconomic Level Via a Multi-Objective Input-Output Approach: A Case Study of the EU-25 Economy

Cortés, D, University Rovira i Virgili
Ruiz -Hernandez, A., University Rovira i Virgili
Guillen-Gosalbez, G., Universitat Politecnica de Catalunya
Llop, M., Universitat Rovira i Virgili
Guimera, R., Universitat Rovira i Virgili
Sales-Pardo, M., Universitat Rovira i Virgili

Originally presented on: 11/4/2013 9:55:00 - 10:10:00

Minimization of the GHG emissions at the macroeconomic level via a multi-objective input-output approach: A case study of the EU-25 economy

ABSTRACT

D. Cortes-Borda[a], A. Ruiz-Hernández[a], G. Guillen-Gosálbez[a], M. Llop[b], R. Guimerà, M. Sales-Pardo.

a.Departament d'Enginyeria Química, Universitat Rovira i Virgili, Av. Països Catalans, 26, Tarragona E-43007, Spain.

b.Centre de Recerca en Economia Industrial i Economia Pública (CREIP), Universitat Rovira i Virgili, Av. Universitat 1, 43204 Reus, Spain.

Global warming has grown significantly in the recent past due to the increasing atmospheric concentration of anthropogenic greenhouse gasses (GHGs). As a result, worldwide environmental policies are placing global warming mitigation as a top priority in the nations' environmental agendas. That is, most of the National governments are adopting policies that aim to minimize GHG emissions.

Unfortunately, in today's globalized markets designing effective environmental policies is challenging because international trade establishes channels through which impacts are imported and exported among nations, overshadowing the main source of pollution into an intricate trading network. Therefore, the design of these policies requires robust environmental assessment methods capable of capturing the complexity of the existing international supply chains of goods/services .

Methods based on extended environmental input-output (EEIO) models are particularly suitable to address global warming at a global scale, since they can assess macroeconomic connections between productive sectors of different nations and the associated environmental impact. This information can be used to allocate the corresponding environmental responsibility. EEIO models are flexible, transparent and accurate enough to perform life cycle assessment (LCA) studies. Hence, LCA-based EEIO models translate the economic output of a given economy into a tangible environmental impact.

By using macroeconomic EEIO models, we can identify a wide range of alternatives for environmental savings, as it is possible to reduce each sector's output (and therefore its associated GHGs emissions) to a different extent through taxation and/or economic penalties. In this context, systematic optimization tools are especially helpful to identify the most promising alternatives considering economic and environmental concerns.

We have developed in this work a systematic approach of this type for reducing the environmental impact of an economy that is based on the combined use of EEIO models and multi-objective linear programming techniques. This tool has been applied to the European Union (EU-25) economy in order to find solutions that simultaneously maximize the total output and minimize the GHGs emissions. Hence, in our approach the problem of identifying key economic sectors that should be given higher priority is posed in mathematical terms as a bi-criteria LP problem that seeks to optimize simultaneously the total European gross domestic product and the associated life cycle GHGs simultaneously. Our study makes use of the E3IOT database, which covers 487 EU economic sectors.

Numerical results show that substantial reductions in global warming potential can be achieved by imposing taxes on sectors such as energy, mineral extraction, and meat and its derived products. By applying this wide-scope assessment, we find potential alternatives for reducing GHG emissions (that are not obvious at first sight) that can assist policy-making.

(36f) Economic Analysis and Optimization of Carbon Capture Storage Pipeline Transport Network System

Dan, S., Seoul National University
Kang, B. J.,
Cho, S.,
Yoon, E. S., Seoul National University
Originally presented on: 11/4/2013 10:10:00 - 10:25:00

Carbon Capture & Sequestration (CCS) is the effective means that can be used to reduce CO2 emissions. Among the CCS network system, a pipeline transport network is the least cost element of the full CCS network system, but the most planning part during the development of the CCS network system.

The research presented here based on the physical properties of CO2 and Co2 containing mixtures calculated. The methodology of this research consists of four steps: Optimal CO2 pipeline design, CCS transport network system modeling, Cost evaluation for model, Selection of the optimal CCS transport network system.

Objective of this research is to find the optimal proposal of CCS pipeline transport network system. This research proposes the optimal engineering CO2pipeline design and CCS pipeline network design for transporting CO2 using HYSYS 7.3 simulator.

(36g) Economic Development Through Entrepreneurship and Technology Innovation: Bringing Together Industry, Academia, and Government

Anid, N. M., New York Institute of Technology

Originally presented on: 11/4/2013 10:25:00 - 10:40:00

The School of Engineering and Computing Sciences (SoECS) of New York Institute of Technology (NYIT) is launching a new Entrepreneurship and Technology Innovation Center (ETIC). The ETIC will consolidate and expand the School's ongoing industry-academic partnerships, foster innovation and promote collaborations between industry, the academic community, professional organizations, and government. The Center will support the NY region's economic development by focusing on three critical areas Health Analytics & Bioengineering, IT & Cybersecurity, and Energy and Green technologies. This paper will discuss a case study of a green economic development initiative that involves chemical engineers: the creation of a microgrid pilot installation that utilizes various storage technologies. Process economics and financing via a public private partnership and seed government funding will be presented. The goal is to infuse the engineering curriculum with real-world problems, and to show how project-based learning, teamwork, and a collaborative environment and direct contact with practitioners, instill in students the key competency and entrepreneurship skills required in today's rapidly evolving, globalized economy.

(37a) The Carbon Capture Simulation Initiative (CCSI): Computational Tools To Accelerate Commercial Development

Miller, D. C., U.S. Department Of Energy, National Energy Technology Laboratory
Originally presented on: 11/4/2013 8:30:00 - 8:55:00

The U.S. Department of Energy's Carbon Capture Simulation Initiative (CCSI) is developing, demonstrating and deploying advanced computational tools and models to accelerate the development of next generation technologies, specifically the development of cost effective carbon capture technologies. This CCSI Toolset will (1) enable promising concepts to be evaluated more quickly based on an optimized process, (2) reduce the time for design and troubleshooting by integrating process and device scale simulations to better predict performance and more effectively resolve issues, (3) quantify technical risk to enable more focused scale up activities, and finally, (4) stabilize the costs during commercial deployment by having a more comprehensive understanding of the process and underlying behavior of the system. This paper will provide an overview of the program and highlight recent accomplishments applying the tools to potential carbon capture technologies.

(37b) The Four Dimensional Product: Integration Over Time Is the Only Way to Understand Sustainability

Banholzer, W., Dow Chemical Company
Jones, M. E., The Dow Chemical Company
Originally presented on: 11/4/2013 8:55:00 - 9:20:00

Sustainability is a question of time, integrating knowledge of production, use and end-of-life. Far too often, hype has surpassed reality, boundary conditions have been neglected and actions taken that don't truly lead to a sustainable future. Singular focus on attributes like renewable content or biodegradability gives the wrong answer. Quality engineering forms the foundation for any meaningful discussion of sustainability, providing insights into energy and material flows, energy return, and thermodynamic constraints. The chemical industry is increasing resource and energy efficiency through innovation in both products and processes. Our mastery of materials science enables unique energy producing and energy saving products, products that improve sustainability by being significantly better than the next best alternatives during use.

(37c) Maximization of Energy Efficiency of a Combined Heat and Power Plant

Edgar, T. F., The University of Texas at Austin
Powell, K., The University of Texas at Austin
Kim, J., The University of Texas at Austin
Kapoor, K., The University of Texas at Austin
Originally presented on: 11/4/2013 9:20:00 - 9:45:00

Energy efficiency is receiving increased attention as a way to reduce the use of fossil fuels and the resulting production of greenhouse gases. Automation, process control, and real-time optimization are key technologies to operate plants in the most efficient way. A combined heat and power plant at the University of Texas at Austin has been studied in order to reduce its carbon footprint as well as reduce costs. The steam power plant has been modeled using steady state mass and energy balances. Chillers have also been modeled so that their efficiencies can be maximized. There is also a thermal storage tank for chilled water to be used during periods of high demand. Forecasting of loads as well as the effect of weather have been incorporated into the real-time optimization approach. The effects of time of day pricing of power and use of demand response techniques to flatten demand curves are discussed. Usage of thermal energy storage along with automatic control gives additional degrees of freedom to deal with dynamic power conditions.

(37d) Pan American Issues on Biofuel Sustainability

Shonnard, D. R., Michigan Technological University

Originally presented on: 11/4/2013 9:45:00 - 10:10:00

Biofuels are in rapid development in many Pan American countries, yet the nature and extent of impacts on social and environmental systems from large-scale biofuel development are not well understood. This talk will explore sustainability issues across a broad set of impacts and in different Pan American countries and will begin to outline important research gaps. A future research agenda to overcome these knowledge gaps would help establish a new and growing biofuels industry onto a sustainable path.

(118a) Sustainability Progress

Abraham, M. A., Youngstown State University

Originally presented on: 11/4/2013 12:30:00 - 13:07:00

Sustainability concepts are critically important in the chemical industry. From a practical perspective, the implementation of sustainability throughout the process industries is increasingly important to our society, and thus to the success of the chemical industry. But sustainability is more than simply green chemistry and engineering, as it involves life cycle and societal implications. Perhaps more so than elsewhere, the production of energy and its use illustrates the importance of sustainability in chemical processes. Interdisciplinary concepts have become the hallmark of sustainable chemical design, synthesis, process and use, and influence sustainable development of existing or new chemicals and materials using a holistic approach. This presentation will describe recent activities in the application of sustainability and sustainable energy as they have been presented over the past few years in the AIChE quarterly *Environmental Progress and Sustainable Energy*.

(118b) Sustainable Biofuels: A Case Of Ethics, Politics, Economics and (I almost forgot) Technology

Sheehan, J. J., NREL

Originally presented on: 11/4/2013 13:07:00 - 13:45:00

The sustainability of biofuels is inextricably linked to the broader challenges of sustainable agriculture and sustainable development. And both these challenges are deeply rooted in political and ethical questions that we have yet to address in a clear and coherent manner. The debate about biofuels' indirect effect on global land use change is the latest and clearest connection between these two problems. Indirect land use change, simply put, considers the problems and possibilities of agricultural expansion and subsequent land clearing associated with increased demands on agricultural land from biofuels.

This land use debate is framed using the "objective" paradigm of the scientific experimenter, which I contend has led us to waste precious research and analysis resources on the wrong question. In this paradigm, biofuel's sustainability rests on a starting assumption that all other aspects of our global agriculture system remain unchanged (the experimental paradigm of ceteris paribus). This is tantamount to asking if biofuels, when added to an already unsustainable agriculture system, has any chance of being sustainable. The answer is obvious: No. But we have spent million of dollars and hours seeking an analytical answer to this question rather than focusing on the questions that really matter. First and foremost, what does it mean for biofuels, agriculture and development writ large to be sustainable? This is as much an ethical question as it is a technical and economic one. In my talk, I will highlight some of the ethical aspects of sustainable development to illustrate how they influence our conclusions about biofuels. Second, what must be done to ensure that global agriculture gets on a path to sustainability? Here, I will discuss a few aspects of modern day global agriculture that offer powerful levers for change. Third, can and should biofuels play a role in sustainable development? The question of "can" largely depends on the answers to the previous question. The question of "should" is a practical and ethical question the answer to which (I will argue) is yes.

(118c) NSF and Support Of Sustainable Engineering Research and Education

Hamilton, B., National Science Foundation

Originally presented on: 11/4/2013 13:45:00 - 14:22:00

An overview of NSF support of sustainable engineering research and education will be presented. Funding opportunities will be highlighted, and brief examples of funded research and education activities will be given. Comments will be made on some potential future directions for research on sustainable engineering. NSF has a wide variety of sustainable engineering research funding opportunities available. At NSF, funding opportunities can be grouped into two general categories, "standing programs" and "solicitations." Standing programs accept proposals on a regular basis, usually at least once per year, during a date window for each program (see the webpages for specific programs on NSF's website,www.nsf.gov). For engineering research on sustainability, NSF's Engineering Directorate (ENG) has established two standing programs that have the word "sustainability" in their program titles. There two programs are the "Environmental Sustainability" program and the "Energy for Sustainability" program. Both of these programs have webpages on NSF's website that describe the programs and how to apply to them for research grants. On each of these program pages, links are embedded that list all the active grants funded by each program, with a description of each grant. In ENG, other programs also accept proposals related to sustainability. For example, the Process and Reaction Engineering program accepts proposals related to the sustainability of chemical processing and bioprocessing. The Civil Infrastructure Systems program will consider proposals on infrastructure sustainability and resilience. The Manufacturing Enterprise Systems program will consider proposals on supply chain sustainability. These are just a few example opportunities in the standing program category. Grants in this category are typically $300K with a duration of 3 years, or, for CAREER grants (Assistant Professors only), $400K with a duration of 5 years. "Solicitations" are special calls-for-proposals, and may be issued only once, or at most, just a few times. For sustainability research, NSF recently established a Foundation-wide investment area, Science, Engineering and Education for Sustainability (SEES). SEES is offering a good number of solicitations, providing diverse opportunities for research proposals. Among these opportunities is the Water Sustainability and Climate solicitation (WSC), with new grant opportunities up to $4 million each. Another solicitation is entitled "Sustainability Research Networks (SRN)." SRN awards are up to $12 million each. One SRN competition has been held so far and resulted in two SRN awards. One of these SRN awards is on the environmental and social impacts of fracking. The other is on climate change mitigation. A second SRN competition is being planned, and should be announced within a few months. Additionally, about 25 grants have been made under SEES to establish Research Coordination Networks (RCN-SEES grants). These are just a few of the activities under SEES, and more information on SEES is available on NSF's website.

(118d) Sustainable Development In The Face Of Water Stress

Reible, D. D., Texas tech University

Originally presented on: 11/4/2013 14:22:00 - 15:00:00

Recent droughts focused attention on the fact that our future is endangered in many parts of the world. Our traditional sources of water are finite. Yet our population will grow and with it our demands for water to support agriculture, economic development and basic human needs. This can only make our economy more sensitive to drought. Moreover, climate trends suggest that water scarcity may become more common and more intense. Without solutions, our growth will ultimately be limited by water scarcity and droughts will have an impact more damaging than any traditional economic recession.

Certainly part of the solution is conservation and demand reduction by all users including industry. It is likely, however, that conservation and demand reduction can at best offset population and economic growth. It is likely that only the implementation of new technologies to advance reuse and recycling and to upgrade poor quality water resources will ensure that water does not ultimately limit economic development. Advanced technologies will be necessary to enable wastes to be treated to an extent to allow greater water reuse and to upgrade poor water sources such as saline groundwaters and produced water. Implementation of these technologies in a manner that will not be equally damaging to the economy, however, will require significant advances in their energy efficiency, reliability and effectiveness.

The presentation will explore these topics, making the case for future needs and discuss some of the challenges and opportunities for technology development to address those needs. The presentation will focus on the application of these approaches to minimize the fresh water requirements of hydraulic fracturing for the extraction of oil and gas from shale.

(198a) Biomass Resources and Current Status Of Biofuels In China

Yuan, Z., Guangzhou Institute of Energy Conversion, Chinese Academy of Sciences

Originally presented on: 11/4/2013 15:15:00 - 16:15:00

With the increase in population and development of industry, energy crisis and environment pollution caused by the fossil energy system have turned out to be two serious problems in the world. In China, the situations appear to be more severe according to the sharp rise in the importation of petroleum in the recent ten years(Fig.1). Biomass is a promising energy source which will be one of the imperative components contributing to the future energy system due to its renewability, cleanness and CO_2 neutral. Laws concerning renewable energy have been issued and some projects were funded in order to promote bioenergy development in China. In this paper, potential biomass resources in China are analyzed(Table 1), current status of biofuels including biodiesel, cellulosic ethanol, liquid alkane and butanol are summarized. Relevant laws and rules are also introduced. Furthermore, development tendencies of biomass energies are predicted.

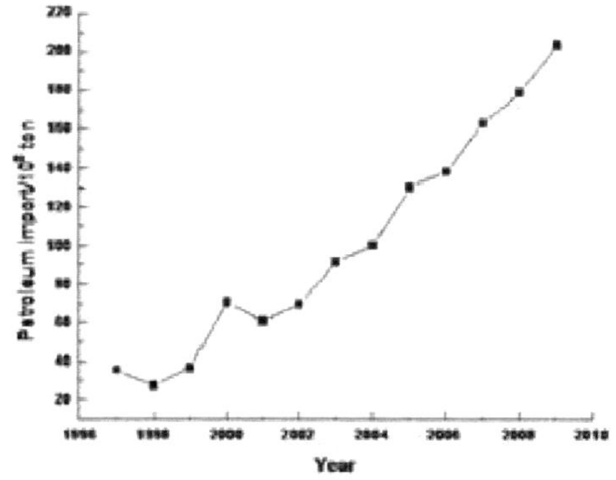

Fig.1 Petroleum importation in China from 1997 to 2009

Table 1. Potential analysis of biomass resources in China (unit:10^8 tons of standard coal)

	2006	2010	2020	2030	2050
Biomass resources in existance	5.0	5.0	5.0	5.0	5.0
In application	2.2	2.0	1.8	1.5	1.0
Available	2.8	3.0	3.2	3.5	4.0
Growth in biomass resources	0.0	0.3	2.3	4.7	5.8
Agricultural and forestry wastes		0.2	1.4	3.0	4.0
Planting of energy crops		0.1	0.4	0.7	0.8
Planting on marginal soils		0.0	0.5	1.0	1.0
Total	**5.0**	**5.3**	**7.3**	**9.7**	**10.8**

(198c) Air Quality Impacts of Natural Gas Production and Use

Allen, D. T., The University of Texas at Austin

Originally presented on: 11/4/2013 17:15:00 - 18:15:00

Natural gas production is increasing dramatically, enabled by technologies such as horizontal drilling and hydraulic fracturing of shale formations. The implications of this transformation in the production of natural gas and associated hydrocarbon liquids will clearly be profound in the energy sector. Much less appreciated, however, is the way in which the widespread availability of natural gas and natural gas liquids will likely transform electricity generation, chemical manufacturing, and agriculture. These changes will have complex environmental footprints. As a case study of these footprints, this presentation will examine air quality impacts associated with natural gas production and use. Data and modeling on emissions and impacts of photochemically active air pollutants, toxic air pollutants and greenhouse gas emissions will be described. In addition to addressing the direct atmospheric impacts of expanded natural gas production, indirect effects will also be described, using a case study of coupled natural gas production and electricity generation. Widespread availability of shale gas, and limited capacities for transporting shale gas to global markets, can drive down natural gas prices, which in turn, can impact the use of natural gas in electricity generation. Natural gas production in Texas and the Texas grid will be used to examine these indirect consequences of expanded natural gas availability.

(198d) Disasters Risks Reduction and Sustainable Development

Robinson, K.,

Originally presented on: 11/4/2013 18:15:00 - 19:15:00

Although the issue of disasters reduction was not included in the Millennium Development Goals (MDGs) framework, it is now evident and widely recognized that disasters, in particular water disasters have made the MDGs attainment more difficult and costly and hampered poverty reduction. They have destroyed human, social and physical capital and derailed social and economic development with long lasting impacts on economic growth. At the same time, the drivers of disasters have not been addressed properly so far. Extreme weather events will increased during the next 15 years and beyond. Within this context, Post-2015 Development Agenda will have to consider the dimension of disasters risks reduction as across-cutting issue that also requires a long-term planning perspective and integration across sectors.

Urban dwellers will increase in numbers and mega-cities are expected to emerge. Particularly vulnerable, however, are the poorest dwellers especially women.

The presentation will address these issues and introduce the United Nations Plan of Action on Disaster Risk Reduction for Resilience including a gender perspective.

(245a) Implementation of Tower Gardens On An Urban University Campus

Peters, D. R. W., University of Alabama at Birmingham
Barbour, N.,
Gaston, S.,
Burns, M.,
Originally presented on: 11/5/2013 8:30:00 - 8:55:00

As society faces the problems of food deserts, diseases of civilization due to malnourishment, and countless environmental harm such as pollution or droughts, creating sustainable communities may serve as a solution to cope with these problems. Implementing green spaces in a form of vertical gardens will have positive environmental effects and offer improved access and quality of fresh produce in the urban as well as rural areas. Growing produce in tower gardens and consuming more unprocessed foods, can potentially reduce rates of undernourishment and many illnesses of longevity.

The Tower Garden® utilizes aeroponics to recycle 100% of its nutrients and water. Furthermore, the system reduces the need for use of pesticides, insecticides, or herbicides. Vertical gardens also use less than 10% of both water and land that are used in traditional farming. They have an undeniable benefit of high yield fresh, nutritious, produce.

Several Tower Gardens are being installed on building rooftops on the University of Alabama at Birmingham (UAB) campus. Raised soil bedded roof gardens are already in-place on several buildings. Knowledge is being gained as to what is most successful in a roof garden in terms of soils and plant materials. Also under study is the growth of plant materials under varying degrees of water and nutrients, and differing locations with respect to sunlight. The use of the Tower Garden will provide the researchers with another system to test and compare with the success of the raised soil bedded gardens already in place. The rooftop gardens promote a sustainable environment as they enhance cooling on rooftops and provide increase insulation for the host building rooftop. This serves to reduce energy use and lowers the building operating costs. The Tower Gardens add two additional benefits to the positive nature of vegetated roofs: (1) the plants grow in an air/water/nutrient mix, with most of the water returned to the atmosphere via evaporation, and (2) the main plant sources are edible produce. For this research project, the Tower Gardens make use of building collected rainwater and equipment condensate as the water source, thereby reducing the storm water runoff from rooftops.

All the benefits of vertical farming contribute to a more sustainable planet.

(245b) New Developments in Implementing Green Engineering in the Chemical Engineering Curriculum

Shonnard, D. R., Michigan Technological University
Allen, D. T., The University of Texas at Austin
Nguyen, N., U.S. Environmental Protection Agency
Austin, S., Chemical Engineering Branch, US EPA
Originally presented on: 11/5/2013 8:55:00 - 9:20:00

The goal of this presentation is to outline updated green engineering and sustainable engineering content for undergraduate and graduate chemical engineering education. The work of chemical engineers is central to the development of sustainable chemical and energy production systems and to meet global human needs. Incorporating environmental issues and sustainability into the design of engineered products, processes, and systems generally means evaluating environmental, economic, and social performance. This presentation will summarize two approaches to incorporate green and sustainable engineering into undergraduate chemical engineering education; **i.** a stand-alone elective course, and **ii.** modules for required core chemical engineering courses. The modular approach, one or more chapters in a forthcoming 2nd edition of "Green Engineering: Environmentally-Conscious Design of Chemical Processes", Pearson-Prentice Hall, can serve as stand-alone supplementary materials for core chemical engineering courses. For example, in a thermodynamics course, environmental equilibrium partitioning and fate properties of chemicals could be a supplemental topic and these properties can be predicted with group contribution or other structure-activity relationships. In a mass balance course, these environmental partitioning and fate properties can be used in conservation of mass problems involving accidental releases of pollutants to various environmental media; air, water, soils, and sediments. These and other examples of incorporating green engineering concepts, methods, and tools in core chemical engineering courses will be presented.

(245c) Incorporating Process Integration Methodologies Into Senior Capstone Design Courses

Dunn, R. F., Polymer and Chemical Technologies, LLC
Debelak, K. A., Vanderbilt University
Originally presented on: 11/5/2013 9:20:00 - 9:45:00

Significant efforts have been devoted to the development of process integration design methodologies over the past several decades. Some of the main developments in process systems analysis using process integration design tools are the identification of heat exchange networks, mass exchange networks and water allocation networks. These systems-based tools are widely applicable and are now essential elements of undergraduate process design instruction. Furthermore, numerous opportunities exist to incorporate these tools in the analysis of senior capstone design projects. In addition, senior design projects based primarily on process integration design methodologies have recently been developed. Experience and lessons learned in applying process integration in senior design projects in the undergraduate curriculum will be presented.

(245d) Assembling Engineering for Sustainable Development Programs

Jensen, C. D., Colorado School of Mines

Originally presented on: 11/5/2013 9:45:00 - 10:10:00

This talk will present four years of action research aimed at developing an Engineering for Sustainable Development (ESD) degree program at a Higher Education Institution known for expertise in mineral extraction, geosciences, materials, and science-technology-engineering-mathematics topics. The action research presented will highlight six stages of research that when incorporated into the education process illustrate the breadth and depth that ESD programs can take. The action research presented continues to evolve and recent outcomes (e.g. a collaborative international summer curriculum) will be presented. This research is a testament that traditional engineering programs can be progressive and make use of holistic concepts and not only reductionist education philosophy or principles.

(297a) A Detailed Mechanism of Solar Thermochemical Carbon Dioxide Splitting With CeO2

Arifin, D., University of Colorado Boulder
McDaniel, A. H., Sandia National Laboratories
Weimer, A. W., University of Colorado at Boulder
Originally presented on: 11/5/2013 12:30:00 - 12:55:00

Currently ceria is a favored material within the concentrated solar power fuels community. The kinetic mechanisms responsible for ceria's catalytic activity have been studied in great detail at lower temperatures and under numerous chemical environments. However, there is a critical gap in the knowledge base of ceria's oxidation kinetics under conditions appropriate for solar thermochemical processes. An understanding of the mechanistic pathway and a detailed engineering model of the oxidation step is necessary to develop accurate computational models for the design and optimization of concentrated solar receiver reactors.

In this work, we focus on ceria mediated CO_2 splitting process as a case study, and apply a solid state numerical approach and a micro-kinetic analysis that separates experimental artifacts from material-specific fuel production behavior; yielding high quality fits to experimentally determined CO production rates over the entire reaction time domain. The results of this study show that the CO_2 splitting reaction is a complex surface phenomenon where temperature-dependent site blocking and mechanism transition occurs at the different oxidation temperatures.

(297b) Isothermal Water Splitting: A Novel Approach to Efficient H2 Generation Through Solar Thermal Energy

Muhich, C. L., University of Colorado at Boulder
Evanko, B. W., University of Colorado at Boulder
Weston, K., University of Colorado at Boulder
Lichty, P., University of Colorado
Liang, X., University of Colorado at Boulder
Musgrave, C. B., University of Colorado, Boulder
Weimer, A. W., University of Colorado at Boulder
Originally presented on: 11/5/2013 12:55:00 - 13:20:00

The use of hydrogen as a renewable fuel has been stymied by our inability to produce it cleanly and economically. Solar thermal water splitting (STWS) cycles have long been recognized as a desirable means of generating sustainable H_2 from water and sunlight. Two-step, metal oxide based STWS cycles generate H_2 by a metal oxide undergoing sequential high temperature reduction and water re-oxidation. The temperatures swings between reduction and oxidation steps long thought necessary for STWS has stifled STWS's overall efficiency and materials robustness due to thermal and time losses during the frequent heating and cooling of the metal oxide. Here, we show that these temperature swings are unnecessary, and that isothermal water splitting (ITWS) exhibits more favorable H_2 production capacity and reaction rates (167.4 μmol H_2/gram and 1.34 μmol H_2/gram/s at 1450 °C) than traditional temperature swing water splitting. ITWS opens the door for more simple, robust and efficient renewable H_2 production. This presentation will summarize the requisite thermodynamics and experimental results for redox water splitting using the "hercynite cycle" .

(297c) Transient Heat and Mass Transfer Analysis of a Porous Reactive Ceria Structure for Solar Fuel Production Via Thermochemical Redox Cycling

Bala Chandran, R., University of Minnesota
Bader, R., University of Minnesota
Lipinski, W., University of Minnesota
Originally presented on: 11/5/2013 13:20:00 - 13:45:00

A high-temperature isothermal solar thermochemical reactor has been proposed for production of synthesis gas from CO_2 and H_2O using ceria-based redox cycles. The reactor cavity is an assembly of annular reactive elements made of highly porous ceria with embedded inlet and outlet gas channels. Simultaneous reduction and oxidation takes place in separate reactive elements. The reactor cavity is subjected to an incident concentrated solar flux of 3000 suns at the aperture. The coupled transport processes of heat and mass transfer along with gas-solid chemical reactions have been numerically modeled in the present study. The developed model is transient, three-dimensional and for a single reactive element undergoing partial reduction when argon gas sweeps through the reactive material. The extended Darcy–Brinkman–Forchheimer formulation is used to model fluid flow through the porous region. For the free fluid domain in the gas channels step-changes are implemented to the values of permeability and porosity allowing the use of a unified momentum transport equation throughout the problem domain. Local thermal non-equilibrium is expected and energy equations are solved in both phases of the porous medium to obtain the respective temperature fields. Rosseland diffusion approximation is employed to compute the internal radiative transport in the solid phase. The incident heat flux on the reactive element is determined by iteratively calculating the net radiative heat exchange inside the cavity. The model is applied to determine influence of geometric design parameters, operating conditions and morphological parameters on the average reaction rate, outlet gas composition and average pressure drop that are treated as indicators of performance of the reactive element. The effect of variations in the thickness of the reactive element, number of elements in the reactor and sweep gas flow rates have been analyzed to optimize the design of the reactive element.

(297d) A Simple, Indirectly-Irradiated, Windowless Reactor for Solar Thermochemical Water Splitting

AuYeung, N., University of Florida
Mishra, R., University of Florida
Greek, B., University of Florida
Martin, B.,
Mehdizadeh, A., University of Florida
Klausner, J., University of Florida
Originally presented on: 11/5/2013 13:45:00 - 14:10:00

Although concentrated solar energy has enormous potential for use as process heat to drive chemical reactions, the thermal efficiency of converting solar energy to chemical fuels must increase by an order of magnitude to become commercially viable. At the University of Florida, a solar thermochemical reactor with potential for high efficiency has been designed and constructed for the purpose of thermochemical water splitting over iron oxide based materials. A 42 kWe solar simulator was used to irradiate up to 10kW of thermal energy through a windowless aperture. Preliminary testing has been carried out to cavity temperatures exceeding 1600°C. Experimentally determined temperature distributions and solar-to-chemical energy conversion efficiencies will be presented. Practical considerations including materials of construction, insulation and sealing strategies will also be discussed.

(297e) Heat and Mass Transfer Analysis of Solar-Driven Steam-Based Gasification of Biochar

Kruesi, M., ETH Zurich
Jovanovic, Z. R., ETH Zurich
Haselbacher, A.,
Steinfeld, A., ETH Zurich
Originally presented on: 11/5/2013 14:10:00 - 14:35:00

Steam-based gasification of biomass using solar process heat has the potential of becoming a low-carbon footprint route to transportation fuels from a renewable resource. At high temperatures and in the presence of steam, biomass is thermochemically converted to syngas (a mixture of H_2 + CO), which can be processed into diesel or kerosene (Fischer-Tropsch process) or first into methanol and then into gasoline (MTG, Mobil). Efficient delivery of the concentrated solar heat to the reaction site is crucial for yield and favorable selectivity of the highly endothermic gasification reactions. The most commonly suggested method for accomplishing such a task is to indirectly irradiate biomass particles entrained through optically opaque tubes with a residence time of less than a second. This, however, requires impractically high operating temperatures as well as energy-intensive and costly preprocessing of the biomass into particles that are small enough to react within such a short residence time. Moreover, due to the low density of the gas-particle suspension the heat transfer to the gas phase is generally insufficient resulting in low gas temperatures, and hence an adverse effect on the selectivity of gas-phase reactions. Alternative concepts such as packed or moving beds allow the use of coarser biomass particles but suffer from significantly reduced overall heat transfer due to high extinction of the radiation by the densely packed opaque solids.

This work investigates a method for increasing the residence time of solids and enhancing heat transfer to the gas phase by incorporating a porous structure into a drop-tube reactor that provides a resistance to the flow of solids. As the particles are gasified while trickling through the porous structure, the mean residence time is longer compared to unimpeded entrained particle flow. Thermal radiation penetrating into the porous structure gets absorbed, thereby enhancing the heat transfer to the gas phase. Because the structure is less optically dense than a packed or moving bed, the penetration depth of the radiation is significantly larger. Based on these enhancements and due to the additional heat transfer by conduction through the structure, a more homogeneous temperature distribution and therefore increased reaction yields are expected.

To explore this concept, a 2-D finite-volume heat- and mass-transfer model was developed. In this model an externally heated tubular reactor is considered, in which activated carbon particles trickle through a porous structure while undergoing steam gasification. The model also includes chemical reaction coupled with conduction, convection, and radiation within the porous structure. The effects of tube diameter, steam concentration, particle loading in the porous structure, and the inner tube-wall temperature were investigated.

The model was validated using steam-gasification experiments with activated carbon particles carried out in an electrically heated tubular reactor. The particles and steam were introduced from the top into the reactor that was filled with the porous structure. The temperature within the reactor and the carbon conversion determined via gas composition measurements were compared to the model predictions.

A more uniform temperature distribution over the tube cross-section for the reactor filled with the porous structure is obtained compared to a packed or moving bed. This is mainly due to the increased solid conductivity but also due to the radiation penetrating further towards the center of the tube. Furthermore, under certain conditions (temperature, tube diameter, particle size, and particle loading) the conversion rate per tube surface area of the proposed configuration is higher than in the packed or moving bed and that higher gas temperatures are achieved than in the entrained flow configuration.

(355a) Design and Optimization of Hybrid Geothermal-Solar Power Systems

Ghasemi, H.,
Mitsos, A., RWTH Aachen University
Originally presented on: 11/5/2013 15:15:00 - 15:40:00

A semi-detailed model is developed for an existing binary geothermal organic Rankine cycle (ORC). The model is
validated with an available set of measured data for one year operation of the ORC. The performance characteristics
of all the components of the ORC is included in the model. The condensation system of the ORC is of ACC type and
consequently the net output power of geothermal system is limited by ambient temperature. The operation of the
ORC is optimized maximizing the net output power of system. The ORC is hybridized with a solar trough system
with two approaches: (i) solar-heating (ii) solar-cooling. A hybrid model for each approach is developed and implemented in AspenPlus. The solar system is of trough medium temperature type and is modeled in Aspen Custom Modeler. The solar system in nominal condition provides only 7.3 % additional thermal energy to the ORC.

In the solar-heating approach, a hybridization strategy is offered and its performance is optimized maximizing the net
output power of the system. The optimal operation of this hybrid system suggests between 10-40 % boost in net
output power compared to the geothermal ORC depending on ambient temperature.

The solar-cooling system is of absorption chiller type with water-LiBr as its working fluid (WF) and no optical concentration. The solar cooling system is utilized in parallel with the ACC system. The optimal operation of the hybrid solar-cooling system offers
up to 4% boost in the net output power of the system.

(355b) Comparison of Potential Alternative Process Options for Solar Energy Storage and Distribution

Muller, B., University of Erlangen-Nuremberg
Muller, K., University of Erlangen-Nuremberg
Arlt, W., University of Erlangen-Nuremberg
Originally presented on: 11/5/2013 15:40:00 - 16:05:00

The TRANS-CSP study conducted by DLR in 2006 concluded that around 630.000 TWh(el) of renewable energy could be harvested economically in desert regions of Northern Africa [1]. However, the envisioned energy generation by Concentrating Solar Power (CSP) plants and its transport by high voltage lines face the drawbacks of low overall efficiency, leading to high system installation cost. Furthermore, for achieving an appropriate annual number of full-load plant operation hours, cost-intensive heat storage devices are necessary.

A new concept for storage and transport of solar energy from the earth´s desert belt is the use of Energy Carrying Compounds (ETS). This concept is based on material pairs showing an energy lean and an energy rich state. Said material pairs can be of inorganic nature, e.g. alkaline earth metal oxides/carbonates that have been presented elsewhere [2]. These ETS materials are converted into each other by uptake of solar energy in an endothermal reaction and release of the same amount of energy in an exothermal reaction. Using materials that are stable under storage and transport conditions, storage cost can be minimized when compared to CSP plants.

Different organic and inorganic material pairs were recently investigated in our group with respect to reaction equilibrium and have been assessed with regard to overall solar-to-electricity efficiency by making use of the process modelling tool Aspen Plus.

In this contribution, material pairs and related process designs for ETS based solar energy storage recently assessed by our group are compared to each other and benchmarked to a reference process using the "classical" concept of solar power generation and distribution by power lines. The work was done within the research framework of Energy Campus Nuremberg (EnCN).

[1] German Aerospace Center (DLR) – Transmediterranean Interconnection for Concentrating Solar Power, 2006, Study commissioned by German government.

[2] B. Müller, M. Helmle, M. Seitz, W. Arlt, Performance assessment of process chains for chemical solar energy storage, 9th European Congress of Chemical Engineering, Den Haag, Netherlands, 2013

(355c) A Numerical Model of Transient Thermal Transport Phenomena in a High-Temperature Solid-Gas Reacting System for CO2 Capture Applications

Yue, L.,
Lipinski, W., University of Minnesota
Originally presented on: 11/5/2013 16:05:00 - 16:30:00

A numerical model coupling transient radiative, convective and conductive heat transfer and mass transfer to chemical kinetics of heterogeneous solid–gas reactions has been developed for a semi-transparent, non-uniform, and non-isothermal particle undergoing cyclic thermochemical transformations. The thermochemical calcination–carbonation reaction pair is selected as the model cycle because of its suitability for solar-driven CO_2capture:

Solar, endothermic step: $CaCO_3 \rightarrow CaO + CO_2$

Non-solar, exothermic step: $CaO + CO_2 \rightarrow CaCO_3$

Mass and energy conservation equations are solved numerically for the solid and gas phases using the finite volume method and the explicit Euler time integration scheme. The solid phase is modeled as an emitting, absorbing and anisotropically scattering porous medium. Rosseland diffusion approximation with non-gray medium radiative properties is used to model the internal radiative transport. A volumetric reaction model is employed for both the calcination and carbonation reactions. Time-periodic boundary conditions for high-flux solar irradiation and the external gas species concentrations are applied. The model predicts the time-dependent temperature distribution, extents of the calcination and carbonation reactions, and consequently the overall amount of CO_2 captured by the single particle over repeated cycles as a function of solar concentration ratio, CO_2 partial pressure in the ambient gas, and particle size. Results support the potential of calcium based sorbent for CO_2 capture, and provide input to solar-driven implementation of the process.

(355d) Solar Thermal Energy-Based High Purity CO2 Release From Carbonate Sorbents

Nakkash, N. B., Al-Nahrain University
Wang, Z., University of Ontario Institute of Technology
Naterer, G. F., Memorial University of Newfoundland
Originally presented on: 11/5/2013 16:30:00 - 16:55:00

Abstract

The increase of the carbon dioxide content in the atmosphere is a major cause of concern in the conservation of nature and in saving the world from natural disaster which follows atmospheric pollution. Reducing CO_2 emissions for addressing climate change becomes increasingly important as CO_2 concentration in the atmosphere has been increasing rapidly, although procedures and measures have been taken to decrease or prevent the increase of carbon dioxide content in the atmosphere. Recycling CO_2 with hydrogen for the production of fuels rather than taking it as a waste is a promising option for our future sustainability development. The subsequent release of the CO_2 after it is captured in a sorbent can be combined with a highly chemical bond energy compound to produce biofuel. This requires a high purity of CO_2 for the fuel synthesis. This paper focuses on the high purity CO_2 release from its sorbents after they capture the CO_2 emissions from an industrial plant.

Thermodynamic analysis is performed in this paper involving the enthalpy changes, the total heat required for the sorbent decomposition at various temperatures, and the maximum heat required by the decomposition with the release kinetics of sorbents are also investigated theoretically.

The present work also concerns with experimentally studying the release of high purity of CO_2 from different carbonates such as $MgCO_3$, $NaHCO_3$, Na2CO$_3$, $CaCO_3$ and $KHCO_3$, using a direct concentrated solar thermal energy. Practically, a solar experimental loop has been used in a solar solarium system to investigate the effects of solar thermal energy and the radiation intensity on the chemical composition profile and the amount of CO_2 released at different temperatures, time intervals, and the total heat required for decomposition of $NaHCO_3$, $KHCO_3$ and $MgCO_3$ to release pure CO2, the results were compared with the theoretical investigations.

(387a) Findings On Life Cycle Aquatic Ecotoxicity Impacts of Ionic Liquids

Mehrkesh, A., University of Colorado Denver
Karunanithi, A. T., University of Colorado Denver
Originally presented on: 11/5/2013 15:15:00 - 15:34:00

Ionic liquids (ILs) are a class of chemicals that are being developed as a replacement for volatile organic solvents for a range of industrial applications (e.g. CO_2 capture, aliphatic/aromatic separations, battery electrolytes, cellulose dissolution). Although ILs can potentially decrease health impacts associated with air emissions due to their nonvolatile nature, many of them are highly soluble in water and can become persistent aquatic pollutants. Environmental impacts of tailored ionic liquids arise both from their direct release into the environment during or after use phase as well as indirect emissions during their synthesis. In this work, we use the USEtox model to quantify aquatic ecotoxicity impacts over the life cycle of an Ionic liquid (IL), 1-butyl-3-methylimidazolium bromide ($BMIM^+ Br^-$), from cradle to grave. The aquatic ecotoxicity impacts from synthesis of this IL as well as its release into the enviorment have been taken into account. Life cycle impact assessment methods define environmental impacts in terms of characterization factors. In this work, for the first time we have developed these characterization factors for ILs. The USEtox, a multimedia fate and transport model, was used to derive fate factors and exposure factors needed for this calculation while a large number of literature data related to toxicity of ILs in a variety of species was used to derive the effect factor. Finally we compare the ecotoxicity production impacts with that of direct exposure.

(387b) Comparative Life Cycle Assessment of Vaccine Cold Chain Infrastructure in Developing Countries

Mousavi, B., University of Colorado Denver
Karunanithi, A. T., University of Colorado Denver
Originally presented on: 11/5/2013 15:34:00 - 15:53:00

An important component of vaccine programs in low-and middle- income countries is the cold chain, which is a collection of cold rooms, refrigerators, freezers and cold boxes that keep vaccines at an appropriate temperature. Vaccines require cold storage (between 2° C and 8° C) in order to remain stable and be potent when administered. Heat sensitive vaccines lose potency after exposure to temperatures above the recommended range and freeze sensitive vaccines lose potency when exposed to freezing temperatures. In the next few years there is an urgent need to streamline and expand efficient and reliable cold chain infrastructure in these countries. In this context, the importance of designing and building new cold chain infrastructure that minimizes the environmental impact of energy, materials and processes both within the country and globally cannot be overstated.

In this study a complete life cycle assessment (LCA) analysis for 5 different types of vaccine refrigerators is done for the case of ghana. All of these refrigerators work in regions where accessibility to grid electricity is very minimal. Firstly, the cradle to grave life cycle inventory for each case and secondly the complete analysis of Global Warming Potential (GWP), Human Health (HH), acidification, ozone depletion and smog impacts are presented. After this LCA analysis, policy makers in developing countries can decide to choose the best option (eco-friendly) based on any environmental impacts.

The 5 different types of vaccine refrigerators selected for this study are:

- solar powered battery supplemented refrigerator (SPBS)
- Solar Powered battery-free refrigerator (SPBF)
- Fully Kerosene refrigerator (FK)
- Hybrid Kerosene- electricity refrigerator (HKE) (biomass cogeneration generates electricity)
- Fully LPG Refrigerator (FL)

The SPBS and SPBF systems work in Vapor-Compression cycle refrigeration system and the other types work in absorption refrigeration cycle. All of these types are specific products from famous companies that produce vaccine refrigerators for developing countries all around the world.

The functional unit of this study was set as providing 1 liter cold storage for vaccines during one year. The life span time for SPBS and SPBF was 15 years and for FK,HK and FL is 5 years.

The cradle to grave unit processes of this study include: refrigerant production, insulation foam production, other materials production, photovoltaic modules production, battery production, assembly of all parts of a refrigerator, transportation, LCI of kerosene and LPG from extraction to combustion and discharge of refrigerant and insulation foam to air at the disposal phase.

The Impact assessment results show that FK and SPBF vaccine refrigerators have respectively the highest and lowest GWP and HH impacts. On the other hand, if the HKE used instead of FK all of the environmental impacts decrease dramatically.

(387c) Hybrid Life Cycle Assessment Model of Silicon Photovoltaics

Yao, Y., Northwestern University
Chang, Y., Northwestern University
Masanet, E., Northwestern University
Originally presented on: 11/5/2013 15:53:00 - 16:12:00

Research on sustainable energy resources is receiving increasing attention because of growing concern about severe pollution associated with fossil-fuel power plants. Solar energy is one of the most promising alternatives. Photovoltaic (PV) technology is gradually becoming competitive in power generation market for its abundant resources and affordable electricity[1]. Among all solar technology, silicon PV technology has been successfully commercialized and has the greatest market share[2]. However, any energy production will also consume energy, generate pollutants and result in some environmental impacts [3]. In order to assess the energy sustainability and environmental impacts of silicon PVs, a lot of work has been done on applying Life Cycle Assessment (LCA) tools to solar energy. Currently, the majority of assessments adopt process-based LCA[1] which is a bottom-up method that is limited by system boundary variations, difficulty for replicating results, and inability to provide assessments for future products due to frequent use of proprietary data. Constrained by incomplete system boundary, results of process-based LCA studies often fail to evaluate the entirety of energy and environmental impacts of solar technology on the whole society and ecosystem. Another tool for LCA studies is Economic Input-output LCA (EIO-LCA) model which has a more comprehensive system boundary, allows for comparisons in systems-level and capacity of assessing future product developments[4]. However, EIO-LCA has time lag for current practices, highly depends on government statistical data, and suffers from sectorial aggregations[5].

In this paper, a hybrid LCA model for silicon PVs, which takes advantage of both process-based method and EIO-LCA, was developed. This hybrid model expanded the system boundary to all supply chain materials of silicon PV by matrix operation (e.g. usage of petroleum coke to produce silicon carbide for wire sawing, electricity generation and usage of all upstream materials). Another strength of this hybrid LCA model is that the geographical shift of PV manufacturing, such as expanding Chinese silicon PV production, can be easily estimated by using publicly accessible input-output data of that country. Since silicon PV has been commercialized for several decades, the variation of technology parameters caused by geographic shift is not significant[1]. However, the difference of sustainability performance of PV made in different countries can still be considerable because of diverse economic structure and energy mix. This difference can be captured and assessed by this hybrid LCA model.

For model demonstration, a case study of U.S. multicrystalline PV production was conducted. Results show that the GHG emissions and energy intensity are 41 g CO_2-eq/kWh (based on 1,800 kWh/m^2/year, 14% module efficiency, 0.75 performance ratio and 30 years lifetime) and 5811 MJ/m^2 (balance of system is included), respectively. The Energy Payback Time (EPBT) is calculated to be 2.54 year. Results of this study are 30 – 60% higher than previous studies using process-based LCA model after harmonization[1]. This is caused by the comprehensive boundary of hybrid model, covering more upstream materials in the supply chain and thus more GHG emissions and energy considerations in research system. Sensitivity analysis was conducted to filter dominant factors influencing the sustainability of multicrystalline PV made in the U.S. from a more comprehensive view, which can provide useful recommendations for future research. Given that this model also shows specific energy consumption and GHG emissions of all sectors in U.S. economy caused by multicrystalline PV production and operation, study findings provide policy maker and general public with a more holistic view of the impacts of PV technology over the whole society and ecosystem. More case studies for silicon PV made in other countries, such as China, will be shown in this paper to give a comparison and show the fluctuation of the sustainability performance of silicon PV caused by geographic shift.

References

[1] D. D. Hsu, P. O'Donoughue, V. Fthenakis, G. A. Heath, H. C. Kim, P. Sawyer, *et al.*, "Life Cycle Greenhouse Gas Emissions of Crystalline Silicon Photovoltaic Electricity Generation," *Journal of Industrial Ecology,* vol. 16, pp. S122-S135, 2012.

[2] "Solar Generation 6," European Photovoltaic Industry Association (EPIA)2011.

[3] EPA, "Annual Energy Outlook 2013," U.S. Energy Information Administration2013.

[4] C. T. Hendrickson, L. B. Lave, and H. S. Matthews, *Environmental Life Cycle Assessment of Goods and Services: An Input-Output Approach*: RFF Press, 2006.

[5] Y. Chang, R. J. Ries, and Y. Wang, "The quantification of the embodied impacts of construction projects on energy, environment, and society based on I–O LCA," *Energy Policy,* vol. 39, pp. 6321-6330, 2011.

(387d) Root-Cause Analysis in the Design of Chemical Processes

Gangadharan, P., Lamar University
Chen, L., Lamar University
Lou, H. H., Lamar University
Originally presented on: 11/5/2013 16:12:00 - 16:31:00

Root-Cause Analysis techniques have found widespread application in quality control, management, failure analysis and process safety. A sustainability root-cause analysis is a powerful tool that will enable designers to identify and display the fundamental causes, related causes and root causes that impact the sustainability of a process. This facilitates the understanding of potential interrelationships between causes which lead to the sustainability issue. A base case in the design of chemical/energy systems generally has high potential for improvement. The challenge is to identify the bottleneck issues that can have a significant impact on overall sustainability.

The Pareto chart and the Fish-bone diagram, in conjunction with the sustainability metrics, form the framework of the sustainability root-cause analysis. The sustainability assessment incorporates economic, environmental, societal, and efficiency criteria. Because of the multi-dimensional feature of sustainability, accounting for the impacts of various design factors and the cause-and-effect relationships can be very difficult. After selecting a process from a group of alternative designs, it is further examined in a root-cause analysis, to identify the major causes that affect the parameters considered in the sustainability assessment, and then eliminate them with appropriate corrective action.

To illustrate the use of the sustainability root-cause analysis tools, several case studies including biodiesel production, polygeneration and natural gas reforming are employed. The Pareto analysis and Fish-bone diagram are invaluable aids to visualize complex relationships and screen out significant factors and correlations for sustainability improvement. This methodology can provide critical guidance to design for sustainability.

(387e) Development of Product-Based Energy Intensity Metrics for the California Food Industry

Walker, M. E., Illinois Institute of Technology
Hutchins, M., UC-Berkeley
Lettieri, D., UC-Berkeley
Romanin, V., UC-Berkeley
Blinde, P.,
Masanet, E., Northwestern University
Originally presented on: 11/5/2013 16:31:00 - 16:50:00

Product-based energy intensity metrics describe the energy footprint of manufactured products. These metrics are valuable for several reasons. From an industry standpoint, determining these metrics are an important first step to the development of an internal energy benchmarking strategy. From a marketing point of view, industrial and home consumers alike are putting an increased focus on green manufacturing and carbon footprints; companies that can highlight their own energy and manufacturing efficiencies may benefit in the marketplace. Finally, in terms of emissions regulation, the development of product-based energy metrics and a product-based benchmarking strategy is the preferred approach.

This presentation focuses on the development of product energy intensity metrics in the California food industry for integration with the state's Cap & Trade program. The development of such metrics for food processing presents a number of challenges, stemming primarily from the fact these plants manufacture multiple products and may lack detailed process monitoring capabilities. This presentation highlights the issues facing the determination of these metrics and discusses the strategies and methodologies applied to develop these metrics in the CA food processing industry.

(387f) Estimations of Direct and Indirect Emissions From Mobile Air Conditioner Units in Automobile Transportation: The Cool Mile Functional Unit

Blowers, P., The University of Arizona
Alayedi, I.,
Dougless, A., U of Arizona
Gerwe, B., U of Arizona
Mesta, A., U of Arizona
Rugel, C., U of Arizona
Wang, Y. Y.,
Originally presented on: 11/5/2013 16:50:00 - 17:09:00

Hydrofluorocarbons (HFCs) may become regulated in the near future by the US Environmental Protection Agency because of their contributions to global climate change. HFCs are currently used as refrigerants and solvents because they are relatively inexpensive and inert. This work focuses on the use of HFCs in automobile air conditioning applications because of the large number of vehicles in use in the US, and the rapid rise of the application of refrigeration in mobile air conditioning systems (MACs) worldwide, with R134a being the predominantly used refrigerant. R-134a (1,1,1,2-tetrafluoroethane,) has emissions of about 38.6 x 10^6 kg/year per year from all sources, with approximately 47% of those due to emissions from vehicles. If R-134a is banned, replacements need to be evaluated in order to determine if greenhouse gas (GHG) emissions are indeed reduced when both direct emissions, in the form of leaks, and indirect ones from changes in fuel mileage, are accounted for.

Papasavva, et al, hafe already analyzed emissions patterns for a new replacement compound, HFc-1234hf, which is a hydrofluoro-olefin from fleet-wide use of the refrigerant in different locations around the world under ambient weather conditions, but did not consider indirect emissions impacts. This refrigerant has also been technically evaluated for use in passenger cars and it appears to be a suitable drop-in replacement. Other refrigerant choices are possible and are investigated in this work, using the "number of cool-miles" as the functional unit in order to capture the changes in emissions due to air conditioning use during drawdown (initially cooling of a vehicle from a hot start) and/or stable maintenance of temperature on longer drives so that the importance of different variables can be explored.

This work shows that some refrigerants have reduced direct emissions due to the refrigerant having a lower global warming potential (GWP), but that indirect emissions due to running the refrigerant through the system sometimes outweigh those direct emissions reductions. Future work should explore how to capitalize on opportunities to minimize the sum of the direct and indirect emissions.

(387g) Regional Air Quality Improvement By Scheduling of Multiple Chemical Plant Start-Ups

Cai, T., Lamar University
Xu, Q., Lamar University
Originally presented on: 11/5/2013 17:09:00 - 17:28:00

Chemical plant concentrated regions may suffer localized and transient air pollution events that violate national ambient air quality standards (NAAQS). Flaring emissions, especially intensive start-up flaring emissions from chemical plants, have potentially significant impacts on local air quality. Thus, when multiple plants in an industrial zone plan to start-up within a same time period, their start-up plans should be evaluated and optimally controlled so as to avoid unexpected air-quality violations in any air-quality concern regions (AQCRs). In this paper, a general systematic methodology for multi-plant start-up emission evaluation and control has been developed. The methodology starts with collecting regional meteorological information such as wind speed and temperature; geographical information of all of the involved chemical plants and AQCRs; as well as plant operation data such as the start-up time window, start-up duration, and estimated emission profile. Next, a regional air-quality evaluation based on Gaussian dispersion model will be conducted. If any air quality violation is predicted to an AQCR, a multi-objective scheduling problem will be generated and solved to optimize the start-up sequence and start-up beginning time for all chemical plants. The scheduling model minimizes the overall air quality impacts to all of AQCRs as well as minimize the total start-up time mismatch of all plants, subject to the principles of atmospheric pollutant dispersion. This study may provide valuable quantitative decision supports for multiple stake holders, including government environmental agency, regional chemical plants, and local communities.

(423a) Incorporating Inherently Safer Designs to Improve Sustainability of Industrial Facilities
Reed, M., MMI Engineering
Peterson, E., Scandpower Inc., a member of the Lloyd's Register Group
Originally presented on: 11/6/2013 8:30:00 - 8:46:00

Incorporating Inherently Safer Designs to Improve Sustainability of Industrial Facilities

Morgan Reed

MMI Engineering

11490 Westheimer Road, Suite 150 Houston, Texas 77077

MReed@MMIEngineering.com

Eric Peterson

MMI Engineering

11490 Westheimer Road, Suite 150 Houston, Texas 77077

EPeterson@MMIEngineering.com

Keywords: Inherent Safety, Sustainability, Process Safety, Process Hazard Analysis, and CFD

Abstract

An increasing number of facilities across the globe are reaching a significant age of maturity. Those facilities were constructed under different regulations and guidelines than those used today. Renovations, upgrades and retrofits are abundant in the oil, gas, and petrochemical facilities and these new constructions must follow the current regulations. The new guidelines will enhance the facility to become more sustainable and inherently safer based on newer design and construction practices. This paper discusses the coming challenges to facilities and the need to incorporate inherently safer designs in the upgrades, modifications to the facilities.

(423b) Designing Greener Plasticizers: Influence of Geometry of Central Group and Side Chains

Erythropel, H. C., McGill University
Shipley, S., McGill University
Brown, T.,
Maric, M., McGill University
Leask, R. L., McGill University
Cooper, D. G., McGill University
Originally presented on: 11/6/2013 8:46:00 - 9:02:00

Plasticizers are important additives to brittle polymers such as poly(vinyl chloride) (PVC). They are mixed with the resin to lower the glass transition temperature (Tg) and their blends result in soft and flexible materials. Yet plasticizers are not bound to the polymer matrix and will slowly leach out of the blend into the environment. Many studies have shown di (2-ethylhexyl phthalate) (DEHP) and some of its stable metabolites such as its monoester mono (2-ethylhexyl) phthalate (MEHP), 2-ethyl hexanol and 2-ethyl hexanoic acid to be toxic. MEHP is under further scrutiny as it is believed to be an endocrine disruptor. The above-mentioned compounds are generally considered ubiquitous environmental contaminants due to their slow breakdown by soil bacteria. This has led to efforts to design new, "green" plasticizers.

Small diester molecules were prepared synthetically by Dean-Stark esterification due to their resemblance to DEHP. These molecules were based on maleic acid (*cis*-isomer), fumaric acid (*trans*-isomer) and succinic acid (saturated analogue), and were esterified with alcohols ranging from ethanol (C2) to n-octanol (C8) as wells as 2-ethyl hexanol. Using a twin screw-extruder, these compounds were incorporated into unplasticized PVC at various concentrations (16% - 33% by weight) and subsequently tested for their plasticizer properties. The Tg reduction was measured using differential scanning calorimetry (DSC). Tensile test bars according to ASTM standards were prepared using a hot press, which were then tested for elongation at break and the secant modulus at a constant strain rate of 5 mm/s. Biodegradation tests were carried out with the pure compounds using the common soil bacterium *Rhodococcus rhodocrous* in shake flasks at 30°C. The medium consisted of minimum mineral salt medium, yeast extract, an auxiliary carbon source hexadecane and the plasticizer. The flasks were inoculated with 1ml of pre-grown culture, and whole shake-flask extractions were performed on a regular basis. The flasks were extracted using chloroform with added internal standard for further analysis by gas chromatography (GC). Using calibration curves for known compounds, plasticizer and metabolite concentrations were calculated. Unknown metabolites such as the respective monoesters were identified by analyzing synthesized standards.

The influence of the central structure was evaluated by comparing 2-ethylhexyl diesters of the proposed molecules to DEHP and another commercial plasticizer: di (2-ethylhexyl) adipate (DEHA). From Tg and tensile strength data, the fumarate compound was eliminated from the list of alternative "green" plasticizers due to poor Tg reduction and poor tensile strength data compared to DEHP. Both di (2-ethylhexyl) maleate and succinate did perform better than DEHP in terms of Tg reduction, and not statistically different in terms of tensile strength.

In the biodegradation experiments a strong influence of the central structure was found: the maleate compound showed almost no hydrolysis of either of the two ester bonds of the molecule over the course of 30 days, as observed similarly with DEHP. In both DEHP and di (2-ethylhexyl) maleate the esters are situated in *cis*-position towards another. The fumarate, in which the two ester bonds are in *trans*-position to one another, an increase in hydrolysis rate was observed, and assuming first-order kinetics, a half-life of 13 days was calculated. For both saturated molecules, the half-life dropped further, to about 6 days in the case of DEHA, and about 2.5 days in the case of the di (2-ethylhexyl) succinate. As expected, there was a buildup of 2-ethanol and 2-ethyl hexanoic acid, however the corresponding monoester was not observed for any of the compounds. In the case of the slow-degrading maleate this is explained by the fact that little of the diester was degraded. It was expected to see some monoester produced by biodegradation of the fumarate and succinate, however, this was not the case. It was concluded that the central structure played a crucial role in degradability of these compounds, and that a saturated molecule that is free to rotate about the central bond is more easily hydrolysed.

Using these results, the succinate compounds were further studied. Instead of the 2-ethylhexyl side chain, even-numbered unbranched alcohols were used to create the ester bond, from ethanol to n-octanol. All four compounds were better at lowering the Tg than DEHP at 29 weight-% of plasticizer in the blend. There was a trend to a lower Tg with increased side chain length, which in turn also meant increased overall molecule length. When the results for dihexyl succinate were compared to the branched di (2-ethylhexyl) succinate, it was found that their Tg at 29 weight-% in PVC were similar, suggesting that the ethyl branch did not play a role in Tg lowering.

Similar to the di (2-ethylhexyl) biodegradation experiments, a rough first-order kinetics approach was chosen to compare relative hydrolysis rates. A trend towards larger half-lives with increasing side chains (less than 1 day for diethyl succinate and about 3 days for di-n-octyl succinate) was observed. When comparing dihexyl succinate to di (2-ethylhexyl) succinate, a difference of about 1 day in half-life was observed, indicating a small steric effect of the ethyl-branch on hydrolysis, however this could also result from a lower solubility of the compound in the aqueous biodegradation broth. None of the succinate diesters with unbranched side chains showed a buildup of any kind of metabolite; in fact it could be shown that the bacteria preferentially grew on the plasticizer or respectively on its metabolites, such as the liberated alcohols. This was shown by running parallel experiments without auxiliary carbon source.

We have demonstrated the potential of small diester molecules to replace DEHP for plasticizing PVC. Given these results, a compromise will have to be found between longer side chains needed to obtain good plasticizer properties, and short side chains for better biodegradability. More mechanical testing and toxicity testing are needed to fully characterize these green plasticizers.

Erythropel, H.C., M. Maric, and D.G. Cooper, *Designing green plasticizers: influence of molecular geometry on biodegradation and plasticization properties.* Chemosphere, 2012. **86**(8): p. 759-66.

Erythropel, H.C., et al., *Designing green plasticizers: Influence of alkyl chain length on biodegradation and plasticization properties of succinate based plasticizers.* Chemosphere, 2013. **91**(3): p. 358-365.

(423c) A Program for Predicting Air Pollution in Chemical, Biological, Radiological, Nuclear and Explosives (CBRNE) Disasters

Lee, K., Yonsei University
Park, J., Yonsei University
Yoon, Y., Inje University
Her, S.,
Moon, I., Yonsei University
Woo, S.,
Yoon, H.,
Originally presented on: 11/6/2013 9:02:00 - 9:18:00

A Program for Predicting Air Pollution in Chemical, Biological, Radiological, Nuclear and Explosives (CBRNE) Disasters

Kwanghee Lee[a], Jaedeuk Park[a], Yi Yoon[b], Sungyun Her[c], Il Moon[a,*]

[a]Department of Chemical and Biomolecular Engineering, Yonsei University, 50 Yonsei-ro, Seodaemun-gu, Seoul 120-749, Korea

[b]Research Institute of Disaster and Emergency Management, Inje University, 197 Inje-ro, Jung-gu, Gyeongsangnam-do 621-749, Korea

[c]National Emergency Management Agency, 209 Sejongdae-ro, Jongro-gu, Seoul 110-760, Korea

Chemical, biological, radiological, nuclear and explosives (CBRNE) disasters involve industrial accidents at chemical and nuclear power plants, incidents from the use of weapons at terrors or wars, and accidents at transportation of hazardous materials. Those disasters incur release of severely hazardous materials into atmosphere, and are featured to be very difficult and expensive to forecast as well as highly lethal once an incident occurs. Internationally representative CBRNE accidents may include those of nuclear power plant at Chernobyl, Ukraine and chemical plant at Bhopal, India, bringing out massive damages to both the accident nation and its neighbors. In recent years, many chemical and radiological accidents have taken place in South Korea, for example, the accident of Hydrogen Fluoride (HF) release at Gu-mi in 2012 killed 5 people and made all Gu-mi residents impossible to sustain their daily business and life. A majority of countries which operate chemical and nuclear power plants have their own emergency response systems. Among them, National Atmospheric Release Advisory Center (NARAC), USA can be thought as being the most advanced worldwide. South Korea has made enormous efforts to develop a set of systems for responsively and effectively managing CBRNE disasters; Chemical Accident Response Information System (CARIS) and Atom-CARE as a system for managing combined chemical and radiological accidents. However, one of the main problems is that the two systems are operated separately in different government ministries, leading to inefficiency in dealing with CBRNE disasters. This study is concerned with the development of a new prototype package for predicting air pollution of combined chemical and radiological accidents, as shown in Figure 1. One of main features of the program is an addition of a new zone taking into account damage on sensitive civilians, namely, "sensitive zone" to existing three zones classified in terms of levels of dispersions of hazardous materials. The prototype package can potentially be expected to be of great help for the development of a program for managing CBRNE disasters in South Korea.

Fig. SEQ Figure * ARABIC 1. Scheme of integrated program

REFERENCES

[1] Michael M, Bradley. NARAC: an emergency response resource for predicting the atmospheric dispersion and assessing the consequences of airborne radionuclides. Journal of Environmental Radioactivity 2007;96:116-21.

[2] Han MH, Kim EH, Suh WT, Hwang WT, Jeong HJ. Development of radiological dose assessment systems to support a radiological emergency preparedness in Korea. Annals of Nuclear Energy 2012;43:187-91.

(423d) Experimental Study and Theoretical Analysis On Decomposition Mechanism of Benzoyl Peroxide

Lv, J., Nanjing University of Science and Technology
Chen, W., Nanjing University of Science and Technology
Chen, L., Nanjing University of Science and Technology
Originally presented on: 11/6/2013 9:18:00 - 9:34:00

Benzoyl peroxide (BPO), a source of radical used as initiator, is an important member in the organic peroxide family. To master its thermal characters, experimental study and theoretical analysis were employed on studying its decomposition mechanism. The autocatalytic process, which was deduced from dynamic experimental data by differential scanning calorimeter (DSC), was testified by isothermal DSC tests. Thermal scanning unit (TSU) was taken to predict pressure hazard of BPO decomposition and accelerating rate calorimeter (ARC) tests were conducted to investigate the thermal behavior of BPO in adiabatic condition. The high pressure and fast pressure release rate indicated a high possibility of accompanied pressure hazard once the runaway decomposition was trigged. Combined with experimental results, quantum chemistry method was used to locate the most possible reaction path and calculate thermodynamic energies of BPO molecule in gas phase. The calculated total released heat during the whole decomposition was in good accordance with the values tested by thermal analysis instruments. And bond dissociation enthalpy (BDE) obtained at different theory levels provided a comparable value in runaway possibility analysis.

(423e) Determination of Ignition Probabilities for LNG Spills

Gopalaswami, N., Mary Kay O' Connor Process Safety Center, Texas A & M University
Originally presented on: 11/6/2013 9:34:00 - 9:50:00

An accidental release of liquefied natural gas (LNG) can lead to the formation of a liquid pool & vapor cloud, which can ignite in the presence of ignition sources. The determination of ignition probability for LNG spills is an important part of quantitative risk analysis to understand the risk people may be exposed to. This paper deals with a review of the models that are currently applied in industry & provides a novel methodology to determine ignition probability. The new ignition probability model takes into account meteorological factors like wind direction, wind class & congestion level, in addition to factors affecting ignition probability which include mass flow rate of flammable materials, its reactivity & the number of ignition sources. An illustrative case study of an LNG setup present at the Brayton Fire Training Field in College Station, Texas is used for calculation of ignition probabilities. A sensitivity analysis is also performed on the new parameters that are involved in the ignition probability calculation. The estimation of ignition probability involves both qualitative identification of sources and its quantification, the results of which can be applied to risk analysis of LNG facilities.

(423f) Zero Valent Silver-Based Electrode for 2, 4-Dinitrotoluene Detection

Mbah, J. C., Tuskegee University
Moorer, K., Tuskegee University
Vahdat, N., Tuskegee University
Hernandez-Rivera, S., University of Puerto Rico
Originally presented on: 11/6/2013 9:50:00 - 10:06:00

Zero Valent Silver-based Electrode for 2, 4-Dinitrotoluene Detection

Mbah J. C.-Tuskegee University

Moorer K. - Tuskegee University

Vahdat N. - Tuskegee University

Hernandez-Rivera S.-University of Puerto Rico-Mayaguez

The prevalent use of nitro-aromatic materials including, 2, 4, 6-trinitrotoluene (TNT) for explosive devices has made their detection crucial in the fight against act of terrorism and detection of military explosives. But rather than attempting to detect TNT, researchers focus on detecting 2, 4-dinitrotoluene (DNT) which is a common impurity in TNT ?based explosive. DNT has a higher vapor pressure than TNT, allowing for more sensitivity and easy detection. In the production of munitions and explosives, DNT is used as a gelatinizing and waterproofing agent. Furthermore, it can be found in landmines, contaminated water, waste sites, and a variety of other locations making it readily accessible to terrorists. High sensitivity and selectivity, combined with the ability to lower the deployment cost of sensors using mass production, is essential in winning the war on explosives-based terrorism.

In this study, the donor based electrode for the detection of DNT was synthesized by modifying the surface of carbon fiber material with zero valent silver (Ag) via chemical deposition. DNT detection was conducted electrochemically in aqueous media of pH 6.0. The cyclic voltammetric oxidation of DNT is diffusion controlled with detection sensitivity increasing with Ag loading of the electrode. The only oxidation product is benzoic acid, which suggests that the difference in performance of Ag-C electrode compared to other available materials is directly ascribable to the electrode fabricated in this study. Detection limit of DNT was 5μM with the electrode showing no signs of degradation after several cycles, and supported by surface behavior of characterized Ag-C electrode. Kinetic study was also investigated. In another study, the fabricated Ag nanoparticle on carbon fiber substrate was used as an efficient surface enhanced Raman spectra (SERS) substrate to detect the adsorbed DNT molecules with a detection limit of 50 ppm. In advent, our developed SERS substrates could have great potential in detecting other nitro-aromatic based-explosive materials, such as TNT molecules.

The objective of this research is to engineer an efficient surface for rapid detection of trace quantities of nitro-aromatic based explosives with particular emphasis on DNT with the following specific aims:

1. Deposit zero valent silver nanoparticles on carbon fiber surface which provide higher surface area-to-volume ratio that can lead to enhanced reactivity with environmental contaminants that degrade through adsorption mechanism.

2. Detect and splits DNT with high selectivity, linearity, and reversibility to a more benign material by electroanalytical means.

3. Modify material surface for surface enhanced Raman spectra with excellent response time and dynamic range.

(423g) Environmental Implications of Phenol Gasification in Supercritical Water

Huelsman, C. M., University of Michigan
Savage, P. E., University of Michigan
Originally presented on: 11/6/2013 10:06:00 - 10:22:00

Supercritical water gasification (SCWG) has been proposed as both a wastewater treatment technology and a biomass conversion technology. Phenol is an important candidate for study in either case, because it is both a regulated EPA priority pollutant commonly found in industrial wastewater effluents and a good model compound for the phenolic structures found in biomass. Additionally, several investigations of SCWG using whole biomass feedstocks identified phenol as a persistent byproduct. Regardless of whether phenol is considered a wastewater pollutant or a byproduct from biomass gasification, our own experiments show that phenol leads to the formation of little gas and a swath of heavier molecular weight compounds when exposed to supercritical water. Most of the non-gaseous products identified in these experiments are polycyclic aromatic hydrocarbons (PAHs), EPA priority pollutants, or have known human health and environmental effects. We experimentally quantified as many of these compounds as possible under various SCWG processing conditions. We then used this experimental data in conjunction with the UNEP/SETAC environmental toxicity ("USEtox") model and other models to characterize the human and ecotoxic impacts due to hypothetical emissions of phenol SCWG byproduct streams into freshwater.

(423h) Modeling the Transport of Mercury in Indoor Spaces Following a Release of Mercury Droplets

Gaglione, A., Eastern Research Group, Inc
Hou, A., Eastern Research Group, Inc
Rindfusz, K., Eastern Research Group, Inc.
Originally presented on: 11/6/2013 10:22:00 - 10:38:00

A mathematical model for predicting potential exposure concentrations of mercury resulting from the breakage of mercury-containing products has been developed. This modeling work estimates airborne mercury concentrations in indoor spaces following the breakage and subsequent release of mercury from mercury-containing products, such as switches, thermostats, and hygrometers. This model was intended to estimate exposure on a per-breakage-event basis. Additional activity factors, such as the frequency of product breakages or the exposure duration to a broken product within these occupational settings, are not included. Two occupational settings are modeled: office and industrial.

This model, named the Mercury Evaporation in Indoor Spaces Model (MEISM), applies engineering principles of liquid-gas phase mass transfer to an industrial hygiene scenario. MEISM models the evaporation of mercury using boundary layer theory and estimates airborne concentrations as a function of time using a near-field – far-field approach to account for imperfect air mixing. The near-field – far-field approach divides the indoor space into two, well-mixed zones and calculates the airborne concentrations in each zone using a mass balance similar to that of a CSTR.

Multiple literature contributions have described the oxidation of mercury in a humid air environment. These studies have noted the exponential decline of airborne mercury concentrations after the initial release. This decline is likely due to the formation of a mercury oxide layer on the surface of the evaporating mercury, which provides a resistance to mercury evaporation. To account for this important phenomenon, MEISM used multiple mercury exposure studies from the literature to regress a first-order rate constant of the oxidation of mercury in humid air. The first-order rate of mercury oxidation is modeled as an exponential decay of mercury surface area available for evaporation. An Arrhenius relationship was used to describe the first-order rate constant as a function of temperature. These regressions resulted in an Arrhenius activation energy of 41.5 kJ/mol and a pre-exponential factor of 578,388 hr^{-1}. This model predicts a first-order rate constant of 0.0312 hr^{-1} at 25 °C.

For the purpose of modeling, MEISM assumes the spilled mercury forms several hemispherical droplets on the floor, and the entire surface area of each hemisphere is available for evaporation. To capture uncertainties of the various model inputs, MEISM uses probabilistic modeling to estimate airborne mercury concentrations following a mercury release. Probability distributions were assigned to the model inputs of temperature, mercury mass within the product, number of mercury droplets formed, indoor air speed, and the oxidation first-order rate constant. For the breakage of a thermostat, switch, and hygrometer in both model office and industrial settings, MEISM predicted the following 95th percentile peak and eight-hour time weighted average (TWA) concentrations in the near field (the zone closest to the mercury source):

Product Type	Mercury Mass Range (g)	Offices		Industrial	
		Near Field 95th Percentile Peak Concentration (µg/m³)	Near Field 95th Percentile 8-hr TWA Concentration (µg/m³)	Near Field 95th Percentile Peak Concentration (µg/m³)	Near Field 95th Percentile 8-hr TWA Concentration (µg/m³)
Thermostats	2.8 – 16.8	7.43	6.55	0.46	0.422
Switches	0.05 – 70	37.2	32.8	2.3	2.13
Hygrometers	3 – 7	4.17	3.67	0.26	0.236

(423i) The Influence of Solid Inertants On the Explosion Behavior of Dusts From Mechanical Waste Treatment Facilities

Held, K., Montanuniversität Leoben

Raupenstrauch, H., Montanuniversität Leoben

Originally presented on: 11/6/2013 10:38:00 - 10:54:00

Conservation of natural resources is a main target in today's lives wherefore the reusing and recycling of waste is very important. To recycle waste material it is necessary to treat the mass flows in special ways. In most instances the waste material passes through mechanical treatments. Especially for the production of substitute fuel mechanical treatment is a common treatment operation. During this mechanical treatment the fractions get shredded, separated and prepared for further recycling. Nevertheless, besides all those positive effects, the mechanical treatment of waste material also generates a fraction of fine particles, which gets dispersed in air. These occurring dusts either have to be separated with suction cleaning or they get deposit unregulated on plant components. Anyway all occurring dusts are fraught with the risk of dust explosions if they are burnable and dispersed in air.

Dusts from different waste treatment facilities were investigated in order to identify the dust explosion potential and its influencing factors. Dusts deposited on plant components as well as filter dusts were sampled. The reason for taking samples from deposit dusts is that they can cause dust explosions if they get dispersed in air. Filter dusts are relevant materials because they are dispersed in air when they get suctioned and might form hazardous explosive atmospheres. Burnable dust dispersed in air can cause dust explosions if the dust cloud is getting in contact with an effective ignition source.

The explosibility characteristics for the investigated dusts were determined by Siwek 20-Liter testing. Experiments were conducted to identify the influence of solid inertants, as well as the influence of their particle size and particle shape. These investigations are done from the intention to predict the explosion behavior of dusts based on the amount, size and shape of their inert fraction.

Because of batch wise delivery of the recycling material and similarly input there is a variation of composition of the treated material over time. That is why it was necessary to check if the occurring dusts respective the explosion behavior of the occurring dusts is altering with time. Beside of short-term fluctuations caused by varying input material also seasonal changes were considered. The comparison of samples taken from the same plant within different periods of time showed that despite of changes in charge material, the explosion behavior of occurring dusts within one operation unit in a plant is reproducible.

Solid Inertants

A known way of reducing the risk of dust explosions is to mix the burnable dust with a solid inertant [1]. This method is either used to generate a resulting mixture, that is rendered non-explosible or the admixture is done in purpose of mitigation of the consequence of a dust explosion [2, 3]. Although there is good knowledge of solid inerting, it is not widely used in process industries. In most applications mixing with solid inertants can lead to unacceptable product contamination so this procedure is only in rare cases practical from a technical perspective.

In the case of dusts from mechanical waste treatment facilities risk reduction due to solid inertants is an interesting approach, because of the fact, that the dusts itself contain a huge amount of inert fraction. The inert fraction is on average at least 40 mass percent.

To identify the influence of inert fraction, dust samples from a mechanical sorting plant were investigated. The ash content of the samples was increased and an inhabitation as shown in Figure 1 could be seen. On this basis it could be shown that the explosion behavior of dust samples from different operation units in the same plant are just differing due to the different inert fractions. So the inert fraction is within one plant a valuable parameter for the prediction of the explosion characteristics of the occurring dusts.

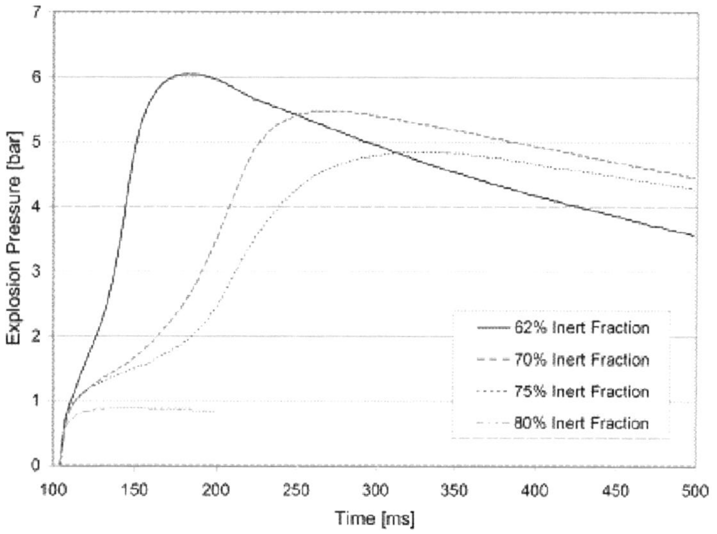

Figure 1: Pressure-time trace for the reaction of dust sample from mechanical sorting with different inert fractions

Particle
Size and Particle Shape

Beside the high amount of inert particles, the fraction of coarse particles in the occurring dusts is very high. Most of the dusts show very wide particle size distributions. For the investigations of the explosion behavior in the 20-Liter Siwek Apparatus as well as in the modified Hartmann Apparatus only particles beyond 500 µm were mentioned. However because of the fact that the particle class of 500 ? 1000 µm is of the same magnitude as the particle class smaller 500 µm, the influence of coarse particles in the mixture was investigated.

An increased mean particle diameter of an explosible dust caused by an increase of particle size over the whole size distribution can significantly decrease the inerting level [1]. Larger particles are more difficult to disperse than smaller particles, which is another relevant aspect of the characterization of dust explosions in mechanical waste treatment plants. Especially, in due consideration of the high LEL (lower explosion limit) of the outlined dust samples a restraint in dispersion properties is considerable.

In Figure 2 the notable results are given. Already at a coarse fraction content of 5 % the explosion characteristics switches to the characteristics of the coarse fraction. Therefore, regardless of separation phenomena, an inhibition of the explosion intensity can be expected because of the presence of coarse particles in the dust samples.

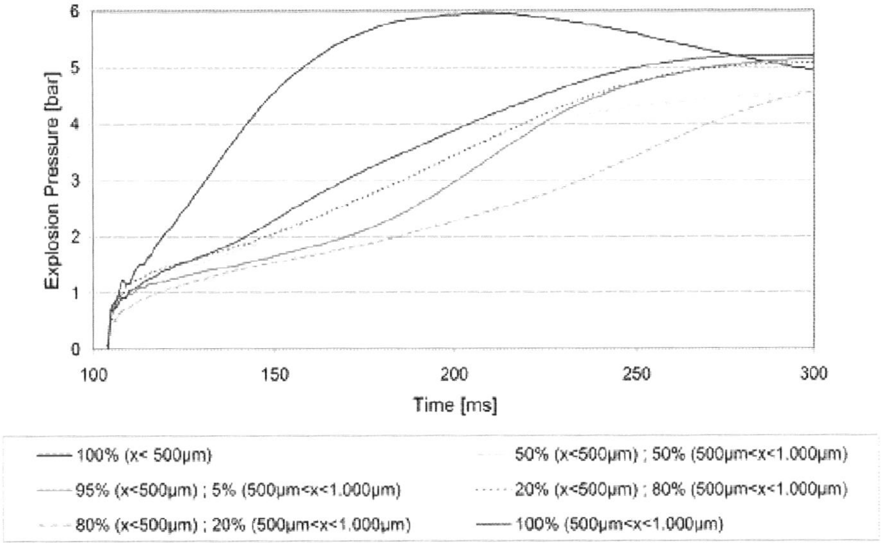

Figure 2: Effect of coarse particles on the explosion characteristics of dust samples from mechanical waste treatment facilities

Particle Size and Particle Shape of Solid Inertants

According to the influence of inert fraction and particle size investigations concerning the particle size and particle shape of solid inertants in the dust samples were done. Therefore a sample from a document destruction plant with an ash content of 42 % was prepared by admixture of glass beads and glass fibers respectively. Due to this the inert fraction of the sample was raised by 5 %.

With an addition of 5 % it could be seen that the effect of beads is nearly negligible while the admixture of fibers has considerable effect. The added materials are of the same equivalent diameter, which is comparable with the particle size distribution of the original sample. Although the explosion pressure of the prepared samples is equal the explosion intensity in terms of rate of pressure rise is strongly weakened (cf. Figure 3). This shows the importance of the particle shape of solid materials that act as inhibitors.

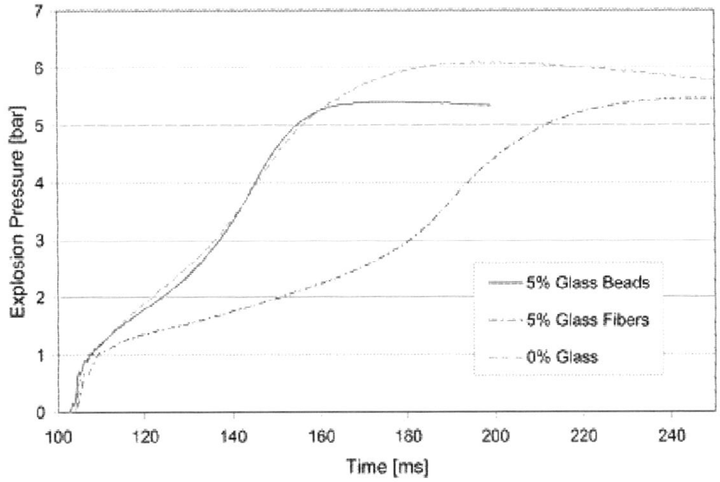

Figure 3: Effect of the admixture of inert material of different particle shape on the pressure rise over time

As seen in Figures 4 and 5 the Siwek 20-Liter testing of the document destruction dust sample with the admixture of 5 % glass beads are compared to those of the same dust sample mixed with 5 % glass fibers. The explosion hazard is therefore significantly magnified by the particle shape of the present inert material.

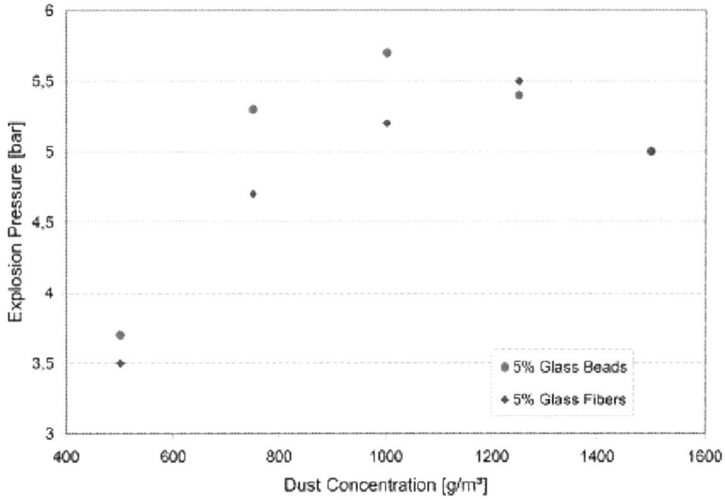

Figure 4: Explosion pressure as a function of dust concentration for dust from document destruction mixed with 5 % glass beads and accordingly 5 % glass fibers

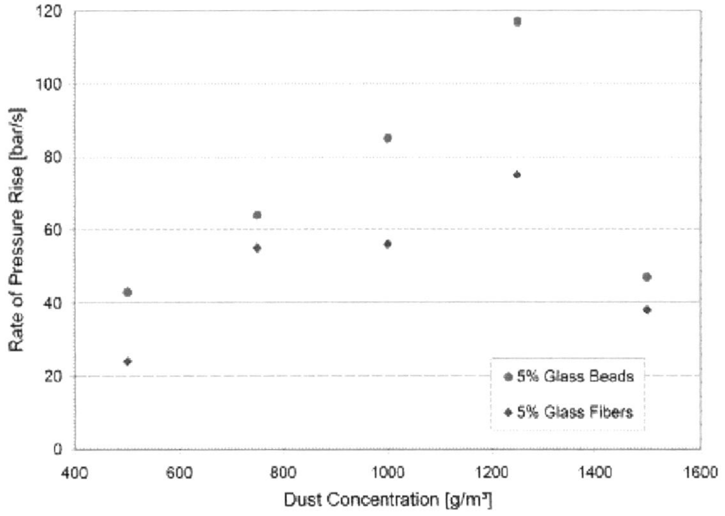

Figure 5: Rate of pressure rise as a function of dust concentration for dust from document destruction mixed with 5 % glass beads and accordingly 5 % glass fibers

These investigations show that the explosion behavior of dusts from mechanical waste treatment facilities is strongly influenced by the presence of inert material and its particle shape as well as its particle size. Previously analyses of different dusts of mechanical waste treatment facilities and the effect of their influencing factors on the explosion reactions were done. This is the basis for further investigations on the effect of particle size and shape of the ash content, which should permit the estimation of the explosion behavior.

Further analyses with variation of the particle size distribution and particle shape as well as the material of the solid inertants will be done. The investigations will be done with dusts from document destruction as well as dusts from substitute fuel production and other mechanical waste treatment facilities. The comparison of the influencing effects provides important information on the characterization of these dust explosions.

References

[1] Amyotte, P., Pegg, M., et al. (2007). Moderation of dust explosions. Loss Prevention in the process industries, 20. 675-687

[2] Chatrathi, K. & Going, J. (2000). Dust deflagration extinction. Process Safety Progress, 19. 146-153

[3] Dastidar, A. & Amyotte, P. (2002). Determination of minimum inerting concentrations for combustible dusts in a laboratory scale chamber. Process Safety and Environmental Protection, 80. 287-297

(444a) Novel Catalysts and Reactor Concepts for Sustainable Processes: the Use of LCA Data in Process Design
Yaseneva, P.,
Lapkin, A., University of Cambridge
Originally presented on: 11/6/2013 8:30:00 - 8:55:00

Novel catalysts and reactor concepts for sustainable processes: the use of LCA data in process design

Polina Yaseneva and Alexei Lapkin

Department of Chemical Engineering and Biotechnology, University of Cambridge, Cambridge, CB2 3RA, United Kingdom

Abstract

We present a multi-step assessment framework, which was developed to guide decision making process in a process development context. Reaction mass intensity metric, gate-to-gate flowsheet analysis and LCA are used at appropriate stage gates. Specifically results of LCA assessment will be discussed for the case studies of a novel catalyst for water purification process, an intensive process of a pharmaceutical API synthesis and a novel organometallic catalytic process for intermediates synthesis.

Introduction

Assessment of sustainability of novel processes for decision-making purposes is a complex task, as frequently uncertainty of the data precludes the use of full LCA methodology, whereas proxy methods such as mass or energy metrics provide incomplete information. At the same time, information about sustainability of novel technologies is now routinely requested even within research projects. Thus, most EU projects in the area of chemical technologies are accompanied by LCA studies.

Within a large integrated project SYNFLOW (www.synflow.eu) we have developed a three-step approach, using reaction mass efficiency metric for selection of chemical routes, gate-to-gate mass and energy metrics for assessment on flow sheet options and finally LCA for evaluation of the demonstration case studies. These steps coincide with stage-gating within a project: confidence in chemistry to be taken up for process development, evaluation of process options with a focus on clean technologies (novel solvents, process intensification) and finally the detailed analysis of complete process at the stage of demonstration with LCA, targeting detailed flowsheet optimisation.

Here we present results of LCA studies for three specific examples. We show LCA of a novel carbon nanotube-based catalyst for water purification, focusing on comparison of the manufacture of the catalyst, a study of a novel flow-process of stoichiometric reduction of artemisinin, leading to manufacture of an important pharmaceutical API (focusing on comparison of batch vs continuous process) and finally a study of a novel catalytic Buchwald-Hartwig amination under overall flow conditions.

Results and Discussion

Carbon nanotube catalyst for water purification

A highly active catalyst was developed for reduction of an emergent aqueous pollutant, bromate, based on 0.3 %wt Pd deposited onto carbon nanotubes, grown on sintered metal fibre. The TOF of the catalyst was found to be an order of magnitude higher than that of a conventional catalyst based on alumina support (Pd/Al_2O_3). However, there remained the question whether high-temperature CVD process used in the manufacture of carbon nanotubes resulted in a cleaner overall process. The LCA study focused on the manufacture of the two catalysts. In this specific case all environmental impacts for the new catalysts were lower. Detailed analysis of contributions of the individual manufacturing steps to various impacts, e.g. analysis of contributions to cumulative energy demand shown in Figure 1, allows to reveal further options for optimisation of the novel catalyst manufacture [1].

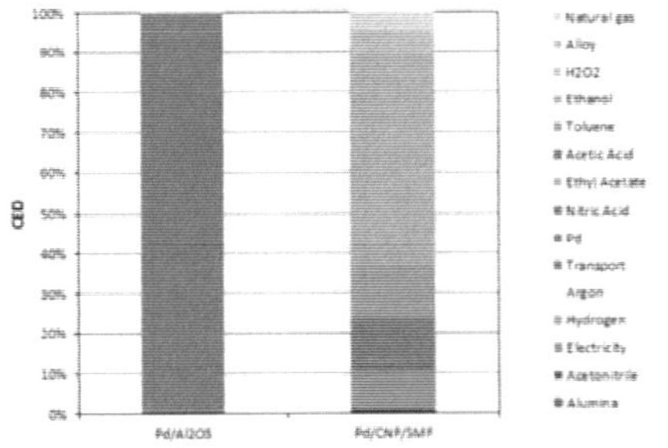

Figure 1. Individual contributions to cumulative energy demand (CED) of the manufacture of the two catalysts.

Artemisinin reduction under intensive flow conditions

We have developed a flow process for reduction of artemisinin to overcome the limitations of a conventional batch process [2]. In the current study we extended the reaction to the final step in the manufacture of the antimalarial API, but also performed an LCA study to compare the novel flow process with the conventional batch reaction [3]. Analysis of LCA results revealed that the solvent used in the manufacture and storage of the reducing agent required for the flow process is having a massive negative impact, see Figure 2. It is becoming critically important to replace the solvent used in a particular stage of the overall process. In this case LCA is instrumental in identifying the critical stage of the overall process that requires optimisation.

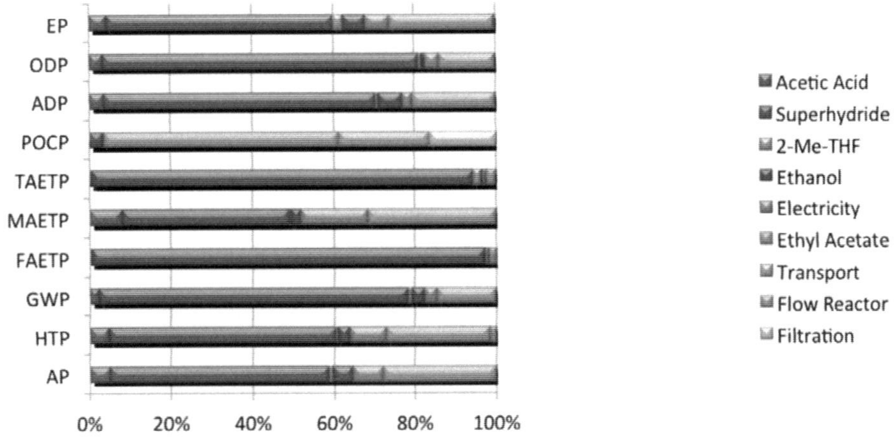

Figure 2. Individual contributions to environmental impacts in artemisinin reduction under flow conditions.

Buchwald-Hartwig amination in flow

The three-step decision making process is illustrated on the example of Buchwald-Hartwig amination. Reaction mass efficiency, process efficiency and LCA are evaluated for this reaction at the different stages in the process design. This approach complements the parallel chemistry-process design methodology being developed within Synflow project, ultimately aiming at a significant reduction of time required for commercialisation of novel technologies, which are also sustainable. Thus, for the initial reaction options we have used RME metric to evaluate the importance of solvent and catalyst recycle. The new process is currently being scaled-up to a demonstrator at an Invite facility of Bayer. We will show the results of stage-gating metrics and initial LCA data.

[1] P. Yaseneva et al., Reduction of bromates on carbon nanofibre supported structured Pd catalysts: experimental and life cycle assessment study, Submitted.

[2] X. Fan, V. Sans, P. Yaseneva, D. Plaza, J.M.J. Williams, A. Lapkin, Facile stoichiometric reductions in flow:

example of artemisinin, Org Proc Res Dev 16 (2012) 1039-1042.

[3] P. Yaseneva et al., Optimisation of the flow process of manufacture of artemether and its LCA study, Submitted.

Acknowledgement: This work was supported by the European Community's Seventh Framework Programme under project "SYNFLOW", NMP2-LA-2010-246461.

(444b) Techno-Economic and Environmental Assessment of Ionic Liquid-Based Biorefineries

Giarola, S., Imperial College London
Shah, N., Imperial College London
Originally presented on: 11/6/2013 8:55:00 - 9:20:00

Although the peak oil is still a debated issue, petroleum supply is expected to become ever more vulnerable of interruptions and current economic system reliance on fossils is exposing all economic sectors to unsustainable price fluctuations [1]. Concerns of maintaining an adequate supply of resources, such as fuels, food, raw materials for a growing population, are also raising a pressing interest for a change in our way to conceive energy and resource provision system [2]. A paradigm shift in the industrial sector from today's large fossils exploitation towards more sustainable practices is also advocated to face current trend of carbon dioxide emissions production, which may soon be approaching greenhouse gas (GHG) levels dangerously close to tipping points of an accelerated global warming [3]. Thus, the future challenge to address the growing world energy demand may be met as long as the supply of higher levels of resources and energy is achieved in a sustainable way promoting low-carbon technologies and renewables [4]. In this respect, the spread of biomass exploitation, as a unique source that can supply both energy and carbon for the production of the commodity chemicals and products, has the potentials to lead to the settlement of a new biomass-based economy [5-6].

Biomass is an abundant, renewable and carbon neutral resource, which is currently significantly exploited for biofuels production from edible components of food crops. However, the major challenge for a future bio-based economy would very likely rely on lignocellulosic feedstock exploitation, being more abundant and less expensive than starchy crops as well as avoiding any direct conflicts with the food chain [7]. Industrial biorefineries have been identified as the most promising route to the creation of a new domestic biobased industry, where biomass conversion processes into fuels, power, and chemicals are integrated through an efficient combination of technologies, to minimise the environmental footprints and ensure the sustainability of all products generated. By producing multiple products, a biorefinery can take advantage of the differences in biomass components and intermediates and maximise the value derived from the biomass feedstock [8]. However, cost-effective technologies for processing lignocellulosic materials into fuels and chemicals are not commercially available. In this context, a central role in determining the overall system profitability is played by the pretreatment unit, where biomass is deconstructed and the three main components (i.e. cellulose, hemicellulose and lignin) are made accessible to subsequent processes [9].

Since the study of Wilkes and Zaworotko [10], ionic liquids have become increasingly popular as reaction and extraction media. The property of ionic liquids which has led to them being classed as green solvents is their very low volatility, thus they do not emit volatile organic compounds and are suitable for reactions at atmospheric pressure. Recently, ionic liquids have begun to be investigated for biomass related processes, yet there are many technological applications and large commercial interests in the field [2]. The use of ionic liquids as lignocellulose deconstruction solvents is also extremely promising [11]. Among the potential applications, ionic liquid can be used to decrystallise cellulose and disrupt the lignin-hemicellulose network. The possibility of removing lignin with ionic liquids and recovering it as precursor for commercial products is also an interesting option [11]. However, deconstruction with ionic liquid will only be viable if some drawbacks are overcome, in particular the cost of ionic liquids. Process design and optimisation is required in order to have high solvent recovery and recycle, but also consumables use reduction, energy integration and waste minimisation. Economic aspects should also be coupled with the concept of ecological sustainability during the process optimisation strategy [12]. Process system engineering techniques play a major role in achieving all the above targets.

This work proposes an investigation of techno-economic performance of a biorefinery based on ionic liquid-pretreatment. The analysis, based on mathematical modelling and process design techniques, aims at identifying the key physical phenomena, the main cost centres affecting the process economics and proposing strategies to develop more efficient biochemical conversion routes for the biomass. This is performed along with the system environmental performance appraisal based on Life Cycle Assessment (LCA) techniques [13]. The final purpose is to provide a mathematical tool for an in-depth economic and environmental analysis of the ionic liquids fractionation, identifying the process enhancements alternatives and outlining potential synthetic routes to biocommodities.

References:

1. Lutz, C., U. Lehr, K. S. Wiebe (2012). Economic effects of peak oil. Energy Policy, 48, 829-834.
2. Tan, S. S. Y., D. R. MacFarlane (2009). Ionic liquids in Biomass Processing. Topics in current chemistry, 290, 311-340.
3. Foley, J. A. (2005). Tipping Points in the Tundra. Science, 310, 627-628.
4. International Energy Agency (2012). World Energy outlook. IEA, Paris (FR).
5. Clark, J. H., F. E. I. Deswarte, editors (2008). Introduction to Chemicals from Biomass Padstow: John Wiley & Sons, Ltd.
6. Ragauskas, A. J., C. K. Williams, B. H. Davison, G. Britovsek, J. Cairney, C. A. Eckert, et al. (2006). The Path Forward for Biofuels and Biomaterials. Science, 311, 484-489.
7. The Royal Society (2008). Sustainable biofuels: prospects and challenges. The Royal Society, London (UK). www.royalsociety.org
8. National Renewable Energy Laboratories (2009). Biomass Program. U.S. Department of Energy, Golden (CO).
9. Aden, A., M. Ruth, K. Ibsen, J. Jechura, K. Neeves, J. Sheehan, et al. (2002). Lignocellulosic biomass to ethanol process design and economics utilizing co-current dilute acid prehydrolysis and enzymatic hydrolysis for corn stover. Technical Report. National Renewable Energy Laboratory (NREL), Golden (CO).
10. Wilkes, J. S., M. J. Zaworotko (1992). Air and water stable 1-ethyl-3methylimidazolium based ionic liquids. Journal of the Chemical Society, Chemical Communications, 965-967.
11. Brandt, A., J. Gräsvik, J. P. Hallett, T. Welton (2013). Deconstruction of lignocellulosic biomass with ionic liquids. Green Chemistry, 15, 550-583.
12. Kralisch, D., D. Reinhardt, G. Kreisel (2007). Implementing objectives of sustainability into ionic liquids research development. Green Chemistry, 9, 1308-1318.
13. International Organisation for Standardisation (1997). ISO 14040: Environmental Management - Life Cycle Assessment - Principles and Framework. International Organisation for Standardisation: (ISO), Geneva.

(444c) Optimal Simultaneous Production of Bio-Ibutene and Bioethanol From Switchgrass

Martin, M., University of Salamanca
Grossmann, I. E., Carnegie Mellon University
Originally presented on: 11/6/2013 9:20:00 - 9:45:00

Ibutene is an important intermediate that has been typically produced from the cracking of the C4 fraction of crude oil. It is the basis for the production of a common additive to gasoline in search for a cleaner burning fuel, the methyl-tert-butyl ether (MTBE), as well as a monomer for the polymerization such as butyl rubber. Its importance can be somehow reflected in its price, around $2/kg. However, the dependency on the crude oil and the already limited availability due to the large number of applications has increased the need for new sources of this chemical. Ibutene has recently been used as intermediate for the production of oxygenated diesel substitutes from glycerol (Chen et al 2011, Vlad et al 2010, Martín & Grossmann 2013). In this application the main drawbacks are, on the one hand, the production of diesel substitutes using a fossil fuel based intermediate and, on the other hand, the economics since from the energy and water consumption standpoint its use for the enhanced production of diesel substitutives is competitive with the process that directly sells the glycerol as byproduct.

In this work, we propose the optimal flowsheet for the production of ibutene from switchgrass. A superstructure embedding a number of alternatives is proposed. Two technologies are considered for switchgrass pretreatment, dilute acid and ammonia fibre explosion (AFEX) so that the structure of the grass is broken down. Surface response models are developed for predicting the effect of the operating conditions on the yield of each pretreatment. Next, enzymatic hydrolysis follows any of the pretreaments to obtain fermentable sugars, mainly xylose and glucose. Ibutene is obtained by fermentation of the sugars. However xylose cannot be easily converted, and thus we also evaluate the possibility of using it to produce bioethanol. Next the ibutene has to be separated mainly from CO_2 for which, PSA or membrane separation are considered while in case ethanol is produced, a multi-effect distillation column followed by molecular sieves are used to dehydrate the ethanol. The problem is formulated as an MINLP with simultaneous optimization and heat integration based on Duran and Grossmann's model (1986). Finally, an economic evaluation is performed.

The most promising process involves the use of dilute acid pretreatment and membrane purification of the ibutene. The production of ibutene using renewable sources is feasible, and may be competitive for its use to generate diesel substitutes for glycerol etherification. However, the decision related to the production of ibutene alone or the simultaneous production of ibutene and ethanol depends on the prices for bioethanol and for switchgrass. For low and current biomass cost the production of ethanol ad ibutene is interesting for bioethanol process above $1/gal.

References:

Chen, P., Min, M., Chen, Y., Wang, L., Li, Y., Chen, Q., Wang, C., Wan, Y., Wang, X., Cheng, Y., Deng, S., Hennessy, K., Lin, X., Liu, Y., Wang, Y., Martinez, B., Ruan, R., 2009. Review of the biological and engineering aspects of algae to fuels approach. Int. J. Agric. Biol. Eng. 2, 1–30.

Duran, M.A.; Grossmann, I.E. (1986) Simultaneous optimization and heat integration of chemical processes. *AIChE, J.,* 32, 123-138

Martín, M., Grossmann, I.E. (2013) Simultaneous dynamic optimization and heat integration for the co-production of diesel substitutes: Biodiesel (FAME & FAEE) and glycerol ethers from algae oil. To be Submitted

Vlad, E., Bildea, C.S. and Bozga, G., 2010, Integrated design and control of glycerol etherification processes, Bull. Inst. Pol. Iasi, LVI (LX), 4, 139-148.

(444d) Synthesis and Design of Biorefinery Processing Networks With Uncertainty and Sustainability Analysis

Cheali, P., Technical University of Denmark
Gernaey, K. V., Technical University of Denmark (DTU)
Sin, G., Technical University of Denmark
Originally presented on: 11/6/2013 9:45:00 - 10:10:00

Synthesis and Design of Biorefinery Processing Networks with Uncertainty and Sustainability analysis

Peam Cheali, Krist V. Gernaey and Gürkan Sin

Department of Chemical and Biochemical Engineering, Søltofts Plads, Building 229, Technical University of Denmark, DK-2800 Lyngby, Denmark

Chemical industries usually rely on fossil based feedstock, which is a limited resource. In view of increasing energy demands and the negative environmental and climate effects related to the use of fossil based fuels, this motivates the development of new and more sustainable technologies for processing renewable feedstocks, with the aim of bridging the gap for fuel, chemical and material production. This project is focusing on biorefinery network design, in particular for early stage design and development studies.

Optimal biorefinery design is a challenging problem. It is a multi-objective decision-making problem not only with respect to technical and economic feasibility but also with respect to environmental impacts, sustainability constraints and limited availability & uncertainties of input data at the early design stage. It is therefore useful to develop a systematic methodology to support the evaluation of processes and the generation of sustainable alternatives for identifying the optimal processing routes. One particular challenge here is to include proactively sustainability analysis as part of the synthesis of biorefinery networks. Another challenge is the handling of several sources of uncertainties such as availability and composition of renewable feedstock, the technical performance of alternative processing technologies and the availability of future markets for biorefinery products, among others.

As part of earlier work in our research group, a systematic methodology to identify/generate optimal biorefineries was developed using the superstructure-based approach, and was implemented in a computer-aided framework. The methodology consists of tools and methods including databases, models, superstructure, and solution strategies to represent, describe and evaluate various combinations of processing networks. The optimization of the network is formulated as a mixed integer nonlinear programming type of problem and solved in GAMS. The methodology was applied for designing optimal biorefinery networks considering biochemical routes. Furthermore, the methodology has also been applied to soybean oil processing networks and industrial wastewater treatment networks. In the work presented here, the methodology for designing optimal biorefinery networks was expanded with inclusion of thermochemical routes, which means the scope and the size of the biorefinery network problem was extended at the level of the database, the models and the superstructure relevant for thermo-chemical conversion routes (the conversion of biomass feedstocks (corn stover, poplar wood) to fuels and chemicals (FT-gasoline, FT-diesel, bioethanol and higher alcohols) via thermal decomposition processes like pyrolysis and gasification).,

In this study, we extend the methodology, models and database by incorporating uncertainty and sustainability analysis as well. Appropriate ranges for uncertain parameters are identified with their correlation/covariance structure and latin hypercube sampling (LHS) is used to sample parameters from their respective domain of uncertainty. The parameter samples are then used as input for solving a deterministic and stochastic optimization problem. The sustainability analysis was performed in two ways: First, it was performed retrospectively to the optimal biorefinery network solution obtained after the MINLP by using an in-house software (SustainPRO) that employs ICHEME sustainability metrics. Secondly, the sustainability analysis was included proactively as part of the MINLP optimization problem that is performed to find the optimal processing path with respect to multi-criteria assessment including technical, economics and sustainability.

The expanded database and superstructure with uncertainty and sustainability analysis form a powerful process synthesis toolbox to be used in design of future biorefineries with multi-criteria evaluation (technical and economic feasibility, environmental impact and sustainability).

(444e) Ecosystems As Unit Operations for Innovative and Sustainable Process Design

Bakshi, B. R., The Ohio State University
Urban, R. A., The Ohio State University
Ziv, G.,

Originally presented on: 11/6/2013 10:10:00 - 10:35:00

Since their development in the 1920s, the concept of unit operations has played an important role in the design of chemical processes through the development of general models and software tools. With increasing importance of reducing the environmental impact of chemical processes, the concept of unit operations expanded to include processes for reducing pollutants in the streams leaving the process boundary. Systems engineering methods based on pinch analysis and optimization have also played an important role in enhancing the efficiency of manufacturing systems. For most chemical processes, it is not practical or economically feasible to completely eliminate material emissions or waste, and the chemical industry continues to be a major producer of hazardous waste and air pollutants, including carbon dioxide.

Ecosystems have the natural ability to treat a large variety of pollutants, and provide goods and services to manufacturing systems. For example, wetlands can provide fresh water by removing contamination by chemicals such as pesticides, oils, phenolic compounds, nitrates, etc. Trees provide water, biomass and oxygen, along with the ability to remove air pollutants such as carbon monoxide, particulate matter, sulfur oxides, carbon dioxide, etc. In fact, an unsustainable system may be defined as a system that produces waste in excess of the natural cleaning capacity of ecosystems. Conversely, sustainability requires that engineering be done within ecological constraints. Unfortunately, most current sustainable engineering methods, including life cycle assessment are not able to satisfy this requirement.

In this presentation, we will “rediscover” the bounty of nature and its importance in sustaining chemical processes. This will involve expanding the concept of unit operations by including ecological processes, to enable engineering design within ecological constraints. Process flowsheets will be designed as networks of technological and ecological systems. Monetary and environmental objectives will be considered, and results for objectives such as maximum profit, zero water, and zero carbon dioxide will be compared. Application to a biodiesel manufacturing process will show that the techno-ecological network is superior to a conventional technology-only process in terms of economic and environmental objectives. Case studies of other chemical processes will demonstrate the need to modify the chemistry or processing methods to permit full benefit of ecosystems. This can lead to new innovation that may not be discovered without including ecosystems among the design options. This approach of techno-ecological networks may be extended to larger spatial scales to result in unique ways of designing and operating supply chains and entire life cycles so that they stay within ecological constraints.

(444f) Process Analysis of Continuous Catalytic Gasification As a Waste to Energy Alternative

Tillie, C. F., Cleveland State University
Riedthaler, M. J., Cleveland State University
Adhikari, S., Cleveland State University
Reeves, S. A., Cleveland State University
Gatica, J. E., Cleveland State University
Originally presented on: 11/6/2013 10:35:00 - 11:00:00

Presently, space exploration waste is either carried onboard for the duration of the mission or discarded into a trash module which is sent back into Earth's atmosphere for destruction. On the other hand, with an estimated cost of $10,000 per pound, the cost of delivering mass to locations beyond low earth orbit (LEO) locations is prohibitive.

Since approximately 15 pounds of waste are produced daily by a crew of four during a mission, the intrinsic value of this waste can be greatly enhanced by waste to energy technologies. Moreover, conceptualizing waste gasification as an in-situ alternative for processing of trash is also an efficient route to control waste, while maintaining a healthy habitable environment, during long-duration missions (Santiago-Maldonado et al., 2010).

The gasification of long-chain polymers is hypothesized to follow a mechanism of four combined reactions —two parallel oxidation reactions which are kinetically controlled, accompanied by two series reaction where oxidation products (carbon dioxide and carbon monoxide) are converted into methane and hydrogen. Integrating this process into spaceflight missions would greatly reduce the onboard mass requirements, as these products can be used as propellant fuel. Further, under certain conditions the process could be also be adapted to produce oxygen and water, two essential life support species for space missions.

This paper explores scale-up alternatives to implement low-temperature continuous gasification processes as potential waste management processes. A kinetic model formulated from batch gasification experiments is integrated into a process simulator along with a detailed thermodynamic description.

Emphasis is placed on the preparatory treatment of the waste slurries, the reactor module, and equilibrium stages for recycling reactants and recovering the catalyst. Detailed characterization of low and mid fidelity simulants is presented.

(445a) Ethyl Acetate Synthesis Using Semi-Batch Reactive Distillation

Kao, Y. L., National Taiwan University
Hsu, L. W.,
Ward, J. D., National Taiwan University
Originally presented on: 11/6/2013 8:30:00 - 8:51:00

Batch reactive distillation (BRD) is an integrated process which combines the advantages of reactive distillation and the flexibility of batch processes. Semi-batch reactive distillation (SBRD) is a design alternative for BRD in which one reactant is fed continuously into the column during the batch. Esterifications of low molecular weight alcohols (methanol–amyl alcohol) with acetic acid have been studied as model systems for the design and optimization of BRD and SBRD processes.

For such esterification processes the lowest-boiling stationary point is an azeotrope. If the azetrope is located in a two-liquid region and the aqueous phase is nearly pure water, (*n*-butanol and amyl alcohol), a BRD process is feasible because water (a product) can be removed from the column by a decanter.[1] Furthermore, Qi and Malone[2] showed that for the more difficult isopropyl acetate system in which the ternary azeotrope lies just inside the liquid-liquid zone, semi-batch reactive distillation (SBRD) operation can improve the production efficiency. Using acetic acid as an entrainer and continuously feeding it into the column during the batch can reduce the concentration of isopropyl alcohol in the distillate so that the distillate composition is closer to the water-acetate edge. Therefore the concentration of water in the aqueous phase is higher.

In this work, the more challenging ethyl acetate process is investigated. Unlike the isopropyl acetate process, BRD with a decanter is not feasible for the ethyl acetate process because the minimum-boiling ternary azeotrope is not located in the two-liquid region. To overcome this problem, researchers have considered the use of a middle-vessel column[3] and a two-column pressure swing BRD process.[4] In this work, we consider another possibility, which is the application of SBRD. Similar to the isopropyl acetate process, acetic acid is fed continuously near the top of the column to reduce the ethanol concentration in the distillate and move the distillate composition into the two liquid-liquid region. After the liquid split, high purity water can removed from the column, and the organic phase containing mainly ethyl acetate and water is collected. High purity of ethyl acetate (99 mol %) can be obtained from the separation of the collected organic product using a batch stripper with a decanter. The two step process is discussed in this work, and some important design variables including the number of reactive trays, initial amount of acetic acid added, side feed policy, reflux ratio policy, and reboiler heating policy are studied to improve the production efficiency.

References

1. Venimadhavan G, Malone MF, Doherty MF. A novel distillate policy for batch reactive distillation with application to the production of butyl acetate. *Ind Eng Chem Res.* Mar 1999;38(3):714-722.
2. Qi W, Malone MF. Semibatch Reactive Distillation for Isopropyl Acetate Synthesis. *Ind Eng Chem Res.* Feb 2 2011;50(3):1272-1277.
3. Steger C, Lukacs T, Rev E, Meyer M, Lelkes Z. A Generic Feasibility Study of Batch Reactive Distillation in Hybrid Configurations. *Aiche Journal.* May 2009;55(5):1185-1199.
4. Modla G. Reactive pressure swing batch distillation by a new double column system. *Computers & Chemical Engineering.* Nov 15 2011;35(11):2401-2410.

(445b) Coupled Permsmbr - Process Design and Development for 1,1-Dibutoxyethane Production

Pereira, C. S., Laboratory of Separation and Reaction Engineering, University of Porto - Faculty of Engineering, Porto, Portugal
Silva, V. T., LSRE - Laboratory of Separation and Reaction Engineering - Faculty of Engineering - University of Porto
Rodrigues, A. E., University of Porto - Faculty of Engineering
Originally presented on: 11/6/2013 8:51:00 - 9:12:00

The PermSMBR is an innovative technology that consists in the integration of permselective membranes with the Simulated Moving Bed Reactor (SMBR)[1]. It was already evaluated for the production of oxygenated compounds, like acetals (1,1-diethoxyethane and 1,1-dibutoxyethane) and a green solvent (ethyl lactate). The PermSMBR shows a productivity enhancement accompanied by a significant reduction on the solvent consumption when compared with other intensified technologies, as reactive distillation and/or SMBR, proving PermSMBR high potential[2-3]. However, in those studies an integrated PermSMBR was considered, which comprises a set of tubular membranes packed with a solid that acts as catalyst and selective adsorbent to water (by-product). In this work, a new configuration is presented, the coupled PermSMBR, where the tubular membranes are located after fixed-bed columns packed with the catalyst/adsorbent. By this way the membranes will not be in contact with the solid, which from an industrial point of view will be easier to implement since the process of membranes installation/replacement and clean-up is simpler. The 1,1-dibutoxyethane production is used, as an example, and each set-up features are discussed. For the studied conditions, the highest productivity achieved using the coupled PermSMBR is 73.44 $Kg_{DBE}.L_{resin}^{-1}.day^{-1}$ with a desorbent consumption of 0.89 $L_{n\text{-}Butanol}/Kg_{DBE}$. The results obtained indicate the coupled PermSMBR suitability for the sustainable 1,1-dibutoxyethane production.

References:

(1) Silva, V. M. T. M.; Pereira, C. S. M.; Rodrigues, A. E. Simulated Moving Bed Membrane Reactor, new hybrid separation process and used thereof. PT Patent 104496; WO Patent 2010/116335, 2009.

(2) Silva, V. M. T. M.; Pereira, C. S. M.; Rodrigues, A. E. PermSMBR—A new hybrid technology: Application on green solvent and biofuel production. *AIChE Journal* **2010,** *57*, 1840-1851.

(3) Pereira, C. S. M.; Silva, V. M. T. M.; Rodrigues, A. E. Green Fuel Production Using the PermSMBR Technology. *Ind. Eng. Chem. Res.* **2011,** *51*, 8928–8938.

Acknowledgement: The research leading to these results has received funding from the European Union Seventh Framework Programme (FP7/2007-2013) under grant agreement n° 241718, EuroBioRef.

(445c) 2-Acetoxy Propionic Acid Using Reactive Distillation â€“ An Intermediate in Synthesis of Bio-Acrylic Acid From Lactic Acid

Kolah, A. K., Michigan State University
Lira, C. T., Michigan State University
Miller, D. J., Michigan State University
Originally presented on: 11/6/2013 9:12:00 - 9:33:00

Acrylic acid, its esters, and acrylate polymers are high volume chemicals and critical chemical building blocks used in a wide gamut of applications including superabsorbents, hard plastics, paints, detergents, and adhesives. The global acrylic acid market is estimated at 4.5 million tons and $12.4 billion annually in 2012, with a growth rate of about 4% per year.

Acrylic acid synthesized from biobased lactic acid is a drop-in replacement for the petroleum-based product synthesized from propylene. Lactic acid is a mature biobased feed stock with a proven track record of high yield (>95%) from glucose at a low cost.

Synthesis of biobased acrylic acid from lactic acid involves a two-step process. In the first step, lactic acid is reacted with acetic acid to form 2-acetoxy propionic acid (2-APA). In the second step of the process, 2-APA is cracked under high temperature to form acrylic acid. The first step reaction is thermodynamically unfavorable, with an equilibrium constant of 0.25, and results in poor conversions if carried out in a batch reactor even with high mole ratios of acetic acid to lactic acid.

In this work, the selective formation of 2-APA by reaction of lactic acid with acetic acid has been carried out using continuous reactive distillation. The reaction takes place in a six-meter tall, 50 mm ID Pyrex column operated at atmospheric pressure, and is catalyzed by Amberlyst-15 cation exchange resin supported inside Sulzer Katapak-SP11 structured packing. Lactic acid feed stocks from 20 wt% to 85 wt% in water were used in experiments along with glacial acetic acid. High conversion of lactic acid and its oligomers were observed (>98%). When 20% lactic acid was used, a selectivity of 96% to 2-APA was obtained.

An unusual observation in the experimental work was the enhancement of the acetic acid – water separation by the presence of lactic acid. Because of this enhancement, it is possible to obtain a distillate stream containing 99% water from the reactive distillation column while retaining acetic acid within the column where reaction occurs.Ternary VLE experiments were performed to obtain the NRTL parameters describing this system.

Experimental studies are complemented by process analysis using AspenPlus simulation software. By implementing a pseudo-homogeneous kinetic model generated in our laboratories from batch experimental kinetic data, and descsribing vapor-liquid equilibria based on experimental data, we are able to predict the experimentally observed pilot-scale column behavior.

It can be concluded that 2-APA can be synthesized very efficiently from aqueous lactic acid solutions by reaction with acetic acid using a continuous reactive distillation column.

(445d) Novel Technology for Producing High Purity n-Hexane and Benzene

Yu, Y., Tianjin University
Li, Y., Tianjin University
Geng, Z., Tianjin University
Ma, J., Tianjin University
Zhang, Z., Tianjin University
Zhang, M., Tianjin University
Originally presented on: 11/6/2013 9:33:00 - 9:54:00

N-hexane is one of the most representative non-polar solvent. It is widely used in pharmaceutical, chemical, polymer materials etc., and owning a great market demand. Benzene is one of the most important petrochemical materials. The yield and production technology of benzene stand for the petrochemical development level of the country. However, the traditional separation and purification technologies, for example, molecular sieve adsorption, batch azeotropic distillation and extractive distillation etc., are difficult to get high yield production of n-hexane and benzene with high purity. Moreover, the energy consumption is high in these processes. Our team has developed a series of new technologies coupling the normal distillation and extractive distillation to produce high purity n-hexane and benzene, and iso-hexane as a by-product. N-methyl pyrrolidone was chosen as the extractant in the process. C3~C8 alkane mixtures from raffinate of aromatics extraction unit, pyrolysis gasoline or straight-run gasoline were used as feed. Gradual separation was used to get the final products. Firstly, the iso-hexane was obtained after removal of the impurities with boiling point lower than 2-methylpentane. Then, the remaining materials were separated by extractive distillation to acquire n-hexane product. Finally, the benzene product was got by further separation. The multi-effect thermal coupling technique was employed in the distillation unit. With the heat integration and optimization, the usage of steam and cooling water was saved. The usage of extractant was also reduced by the way of the cycling between the extractive distillation column and the recovery column. To prevent impurities accumulated in the extractant, removal of the heavy component impurities was implemented by taking a small part of material into the falling film evaporation system. High purity n-hexane and benzene product were finally produced by the above mentioned method. The purity of n-hexane product is higher than 99 wt% with the content of benzene is lower than 100 ppm. At the same time, the concentration of benzene product is higher than 99.9 wt%. Currently, a 30,000 MTA n-hexane and benzene plant has been built up using this technology.

(445e) Online Monitoring of Emulsion Polymerisations Using Raman Spectroscopy

Houben, C., University of Cambridge
Nurumbetov, G., University of Warwick
Haddleton, D.,
Lapkin, A., University of Cambridge
Originally presented on: 11/6/2013 9:54:00 - 10:15:00

Many polymer products are manufactured using batch and semi-batch reactors. Typically the process parameters are fixed and are not varied during the reaction. Unexpected changes in operating conditions can lead to variations in polymer structure and purity, resulting in significant losses due to off-spec materials. An approach addressing the complex issues of real-time process control of large-scale emulsion polymerisation processes is to use advanced models and on-line sensors. A specific objective is to implement a model-based predictive control based on Raman spectroscopy.

(445f) Process Intensification in Millistructured Extraction Columns

Holbach, A.,
Caliskan, E., TU Dortmund University
Kockmann, N., TU Dortmund University
Originally presented on: 11/6/2013 10:15:00 - 10:36:00

Process intensification in millistructured extraction columns

The downstream process of biochemical and pharmaceutical products can account up to 80% of the cost of production. [1,2] To reduce the cost it is necessary to develop new process intensified equipment with higher time-volume-yield. Process intensification is one important part to make the industry more economical and ecological. Some mass or heat transport limited chemical reactions are only technical possible or cost-efficient with process intensified equipment. The small channels and structures in this equipment gives the potential to control the fluids on the sub-millimeter scale. The known characteristics of transport phenomena enable an easy and fast scale up for bulk chemicals from lab to industrial scale.

We investigated the process intensification in millistructured extraction columns with an inner diameter of 15 mm, see figure 1. The column has 20 stirred cells and an active extraction high of 440 mm. The total internal volume with head and bottom is 220 ml with total height of 760 mm. With stirred cells we achieved high energy input, high specific surface between the two phases and thus good mass transfer. The backflow was decreased with narrow plates between the plates. Thus, the dispersed phase coalesces at the plates and plugs the stirred cells. With axial pulsation of the fluid the coalescence area is destroyed and the dispersed phase is transported into the next cell. This new technique allows for controlled flow of the dispersed phase into the next stage. The retention times for counter current processes can be adjusted for transport limited processes or reactive extraction systems.

The hydrodynamics and mass transfer inside the column was characterized with the European Federated Chemical Engineering (EFCE) testsystem n-butylacetate(d)/acetone/water(c). The mass transfer from continuous-to-disperse (c-d) and from dispers-to-continuous (d-c) and the droplet size distribution with extraction efficiency was investigated. Different to conventional columns, the energy input inside the stirred cells are higher and the influence on the coalescenc and mass transport direction is not so large. We achieved specific surfaces between the two fluids up to 3200 m^2/m^3 and measured the mass transfer up to 7 theoretical extraction stages per active high (H = 440 mm). This correspondsto 16 theoretical stages per meter, see figure 1. The experiments show that higher total throughput yields higher extraction efficiencies, hence, high mass flow rates are preferred.

The enantioselective extraction of (R/S)-α-cyclohexyl-mandelic acid will be investigated as an example; see the molecules in figure 2. The enantioselective extraction is an organic/aqueous reactive extraction. The enantiomers create in the aqueous phase a complex with hydroxy-propyl-ß-cyclodextrin. The (S)-complex is preferred to the (R)-complex, so that the extraction is selective. The whole process is mass transport limited and has a low selectivity. In conventional columns the extraction is impossible, because the retention time of the dispersed phase and the specific surface between the two fluids are too small. In process intensified extraction columns with high specific surfaces and many extraction stages the counter current extraction is possible. First results show a considerable separation of the two enantiomers.

[1] Lowe C. R., Lowe A. R., Gupta G., New developments in affinity chromatography with potential application in the production of biopharmaceuticals, J. Biochem. Biophys. Methods 49, 2001, 561?574

[2] Strube J., Sommerfeld S., Challenges in biotechnology production?generic processes and process optimization for monoclonal antibodies, Chemical Engineering and Processing, 44, (10), 2005, 1123-1137

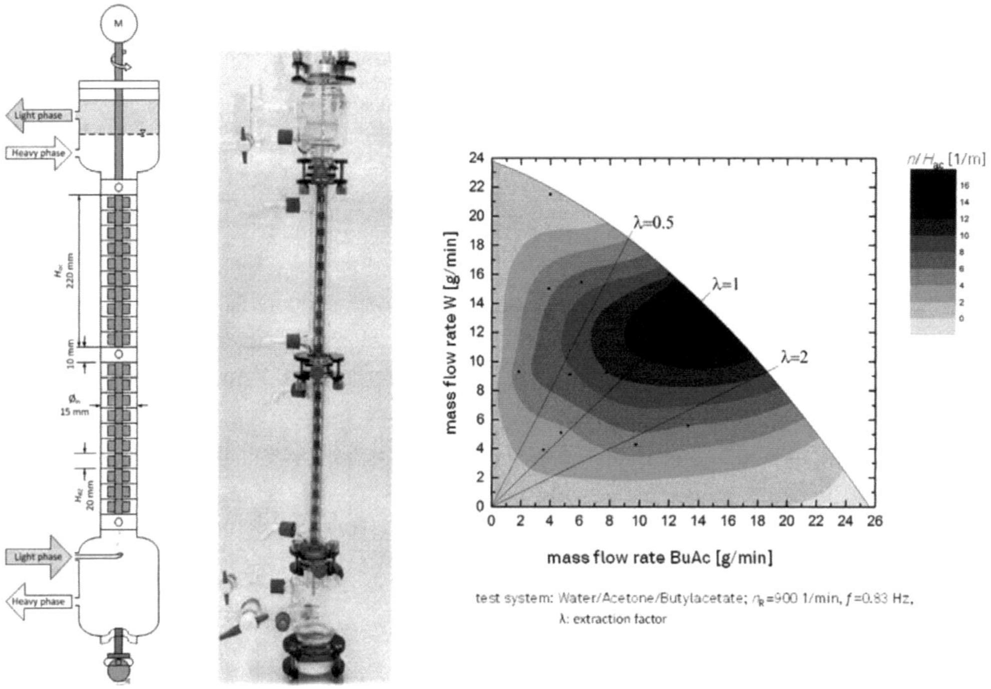

Figure 1: Millistructured extraction column; left side: sketch and picture of the column with two sections; right side: experimental extraction efficiency with different mass flow rates

(S)-α-cyclohexyl-madelic acid

(R)-α-cyclohexyl-madelic acid

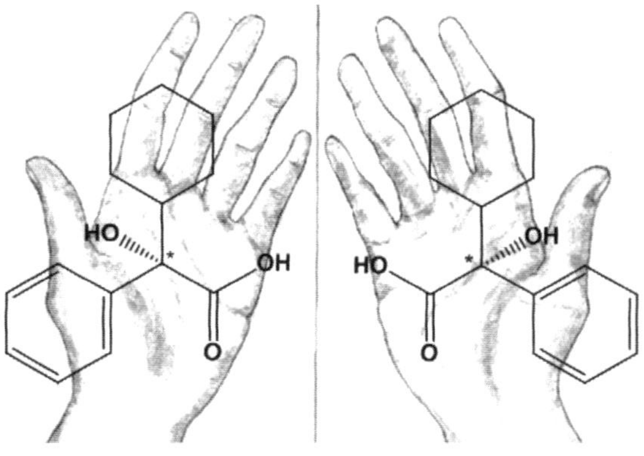

(S)-α-cyclohexyl-madelic acid (R)-α-cyclohexyl-madelic acid

Figure 2: Molecular structure of (S) and (R)- α-cyclohexyl-madelic acid

(445g) A NOVEL Comparison Among Intensified and Traditional Complete Ethyl Acetate Productions Designs

Santaella, M., Universidad Nacional de Colombia
Narvaez, P. C., Universidad Nacional de Colombia
Orjuela, A., Universidad Nacional de Colombia
Originally presented on: 11/6/2013 10:36:00 - 10:57:00

Over the last decade different proposals of process intensification have been made for the improvement of the esterification reaction between acetic acid and ethanol. The dissimilar considerations and in some cases unrealistic assumptions made by the authors during modelling limit the possibility to compare among different production processes. In addition to this, different inlet compositions and different product specifications represent a major barrier to ensure an objective comparison between different processing technologies.

In this paper, the conventional production process, the reactive distillation (RD) and the reactive dividing wall column (RDWCD) process for the production of urethane grade ethyl acetate (99,5%) have been simulated using a commercial software (Aspen One 7.2). Glacial acetic acid and azeotropic ethanol were used as raw materials, and different catalysts were evaluated (self-catalytic, sulphuric acid, Amberlyst 15, Amberlyst 70). A validated NRTL-HOC thermodynamic model was used to describe phase equilibria. Chemical kinetic models were obtained from literature. A preliminary evaluation of reactive residue curve maps obtained with different catalysts and residence time ranges between 0-500min allowed to compare feasible column profiles.

Capital costs and energy savings for reactive distillation are predicted between 15-80% over classical configurations of a reactor followed by separation units. Dividing wall column, Petluk column and similar thermally coupled configurations predict 30-50% energy savings over conventional two distillation column processes. The RDWCD also predict 30-50% energy reduction over simple RD.

(492a) Industrial Boiler Systems and the Energy-Water Nexus

Walker, M. E., Illinois Institute of Technology
Masanet, E., Northwestern University
Originally presented on: 11/6/2013 12:30:00 - 12:52:00

Industrial systems consume a great deal of water, a large portion of which is utilized in facility boiler systems. As a first step towards reducing freshwater use and improving manufacturing sustainability, it is important to develop a firm understanding of the breakdown and magnitude of water consumption in these systems. However, comprehensive water use data is not available on an industry-by-industry basis, and is particularly difficult to find on an end-use basis, (cooling water, boiler makeup, etc.). This is in contrast to energy use data, which is collected on a relatively rigorous basis, at a detailed level, and is publicly available. This presentation will discuss the application of a thermodynamic engineering model to a boiler system to obtain previously unavailable information on industry-level water use for industrial boiler systems. This model was applied to detailed energy use information from the 2006 Manufacturing Energy Consumption Survey (MECS) in combination with operation data for the 60 largest energy consuming manufacturing industries. Results show that an estimated 370 million gallons of freshwater is consumed by industrial boiler systems throughout the U.S. every day; approximately 350 MGD of which is attributable to paper manufacturing, chemicals production, petroleum refining and food processing. This presentation will also highlight estimates of total potential and realized water savings in over 1,000 major U.S. manufacturing plants, based on energy savings audit data for steam systems at these same plants.

(492b) Optimization for Sustainable Scheduling of Batch Manufacturing Processes Under Economic and Environmental Concerns

Yue, D., Northwestern University
You, F., Northwestern University
Originally presented on: 11/6/2013 12:52:00 - 13:14:00

Batch processing plays an important role in producing low-volume high-value-added products due to its flexibility, thus the subject of batch production scheduling has been extensively studied in the past decades, and still remains an active research area nowadays both in industry and academia [1, 2]. While most works tend to improve the economic performances of batch scheduling, the associated environmental impact has long been disregarded, which becomes the focus of this work. Due to the increasingly stringent environmental regulations imposed on chemical process industries and the call for green manufacturing from the public, it has become a goal in process manufacturing to incorporate the environmental consideration in addition to the economic objective in the decision-making process [3].

In this work, we propose a novel bi-criterion optimization framework which simultaneously takes into account the economic and environmental concerns in batch scheduling problems [4]. Two fractional-term objective functions are considered that allows further improvement compared to the conventional linear economic and environmental objectives. The economic performance is measured by productivity, defined as the net profit divided by the corresponding makespan. This objective would take into account both the absolute profitability of given batch processes as well as the temporal responsiveness perspective. The environmental performance is measured by the environmental impact embedded in per functional unit of the products, evaluated through a "cradle-to-gate" Life Cycle Assessment (LCA) procedure [5, 6]. We explicitly optimize based on the functional unit in this work. Moreover, the interpretation phase of classical LCA is coupled with the multi-objective optimization, thus leading to the concept of Life Cycle Optimization (LCO). All decisions (e.g., batch sequencing, task assignment, processing timings, batch sizing, etc.) are simultaneously optimized under both the economic and environmental objectives. Furthermore, it worth highlighting that the selections among multiple changeover and cleaning options with different economic and environmental properties are considered as variables in this work, which is rarely considered in conventional batch scheduling problems.

The bi-criterion optimization model is solved using the ε-constraint method, in which the Pareto-optimal solutions are derived by solving a sequence of subproblems with different ε-parameter values. Due to the presence of fractional economic and environmental objective functions, the subproblems are formulated as mixed-integer linear fractional programming (MILFP) problems. MILFP is a special class of mixed-integer nonlinear programming (MINLP) problems which can become computationally intractable due to its combinatorial nature and pseudo-convex objective functions. Therefore, to globally optimize the MILFP subproblems, we employ two tailored solution methods, namely the parametric algorithm and the reformulation-linearization method, which are proved to be much more efficient than general-purpose MINLP methods [7, 8]. The resulting Pareto curve reveals how the inherent tradeoffs between the economic and environmental objectives would influence the production schedule and resource allocation in batch manufacturing, which provides insights to the sustainable scheduling problem. To illustrate the application of the proposed modeling framework and solution methods, we consider two short-term sustainable scheduling problems as our case studies, which involve a multi-product acrylic fibers batch production process and a multi-purpose batch plant with a reaction-separation network structure, respectively.

References

[1] C. A. Mendez, J. Cerda, I. E. Grossmann, I. Harjunkoski, and M. Fahl, "State-of-the-art review of optimization methods for short-term scheduling of batch processes," *Computers & Chemical Engineering,* vol. 30, pp. 913-946, May 2006.

[2] C. A. Floudas and X. X. Lin, "Continuous-time versus discrete-time approaches for scheduling of chemical processes: a review," *Computers & Chemical Engineering,* vol. 28, pp. 2109-2129, Oct 2004.

[3] E. Capón-García, A. D. Bojarski, A. Espuña, and L. Puigjaner, "Multiobjective Optimization of Multiproduct Batch Plants Scheduling Under Environmental and Economic Concerns," *AIChE Journal,* vol. 57, pp. 2766-2782, Oct 2011.

[4] D. Yue and F. You, "Sustainable scheduling of batch processes under economic and environmental criteria with MINLP models and algorithms," *Computers & Chemical Engineering,* vol. 54, pp. 44-59, 7/11/ 2013.

[5] A. Azapagic, "Life cycle assessment and its application to process selection, design and optimisation," *Chemical Engineering Journal,* vol. 73, pp. 1-21, Apr 1999.

[6] A. Azapagic and R. Clift, "The application of life cycle assessment to process optimisation," *Computers & Chemical Engineering,* vol. 23, pp. 1509-1526, Dec 1999.

[7] F. Q. You, P. M. Castro, and I. E. Grossmann, "Dinkelbach's Algorithm as An Efficient Method to Solve A Class of MINLP Models for Large-Scale Cyclic Scheduling Problems," *Computers & Chemical Engineering,* vol. 33, pp. 1879-1889, Nov 2009.

[8] D. Yue, G. GuillÃ©n-Gosálbez, and F. You, "Global Optimization of Large-Scale Mixed-Integer Linear Fractional Programming Problems: A Reformulation-Linearization Method and Process Scheduling Applications," *AIChE Journal, submitted,* 2013.

(492c) Modeling and Multi-Objective Optimization for Robust Design of Sulfuric Acid Purification Process in Semiconductor Industry

Song, K., School of Chemical and Biological Engineering, Seoul National University
Jung, J., School of Chemical and Biological Engineering, Seoul National University
Park, S., Seoul National University
Han, C., Seoul National University
Originally presented on: 11/6/2013 13:14:00 - 13:36:00

Sulfuric acid is a widely used solvent in semiconductor manufacturing for cleaning the wafer. Pure and highly concentrated (usually higher than 80 wt%) sulfuric acid is consumed to effectively eliminate remaining particles in the wafer. Waste acid is discarded after dilution or repeatedly used after purification treatment. However, tighter environmental regulations promote on-site acid recovery system.

Main design issues of the system in the semiconductor industry are the following. First, semiconductor grade materials should be used for stable repetition of acid recovery. This results in high capital investment requirement. Second, the equipment is recommended to be designed to be compact and in module type for on-site installation. Third, the recovery system is required to handle various waste acid feed conditions resulting from different end-users. This means that the device should be able to deal with consumed sulfuric acid in various temperature and concentration.

In this work, authors present modeling and optimal design of sulfuric acid purification process. Vacuum distillation in batch process consisting of tanks, a condenser and a vacuum pump is designed to produce concentrated sulfuric acid and segregate waste water. Vacuum condition is essential to meet operating temperature limit posed by material selection. Dynamic modeling of batch process is performed to effectively calculate operating time and steam generation rate, which directly defines the condenser sizing. An optimal design in terms of minimizing total capital cost and operating cost is deduced first. Economic evaluation and full discussion of sensitivity analysis in main variables is performed. Then, robust design with maximizing the operating range to cope with various feed conditions as the objective function is solved in a multi-objective formulation.

(492d) Multistage Technology Integration for Sustainable Manufacturing of Complex Industrial Systems

Yan, L., Department of Chemical Engineering and Materials Science, Wayne State University
Huang, Y., Wayne State University
Originally presented on: 11/6/2013 13:36:00 - 13:58:00

Numerous technologies are available for process efficiency improvement, product quality control, waste reduction, safety assurance, etc. However, how to identify the most appropriate technologies for cost-effective, stage-wised sustainability enhancement in industrial systems is always a challenge. It becomes more complex if an industrial organization needs to evaluate the feasibility of different technology integration strategies and explore the pathways towards different sustainability goals in multiple development stages. It becomes even more challenging when the available technical and nontechnical information is uncertain and incomplete.

In this paper, we introduce a multistage sustainability enhancement methodology, which can be used to formulate an industrial sustainability problem as a multi-objective optimization problem and to identify a set of ranked solutions for different goal achievement stages. A unique feature of the methodology is its capability of deriving optimal solutions using uncertain information. A sophisticated distributed biodiesel manufacturing problem is studied, where technology integration options are evaluated based on regional feedstock availability, fuel demands, transportation infrastructure, economic development plans, environmental regulations, etc. The methodology is general so that it is applicable for the study various industrial sustainability problems.

(492e) Green Manufacturing of Sustainable Self Heating Products

Sunol, A. K., University of South Florida
Sunol, S. G., University of South Florida
Li, H., University of South Florida
Smeltzer, B., University of South Florida
Cogswell, K., University of South Florida
Originally presented on: 11/6/2013 13:58:00 - 14:20:00

Several self heating and cooling products that use novel green compositions are designed using finite element methods, rapidly prototyped in part with the aid of 3D printing, and sustainablly manufactured. The supply chain are optimized for two of the sustained release self heating products that have been recently introduced into market. Unique chemistry displace products that utilize security risk hydrogen releasing chemical compositions. The smart coating polymers with green encapsulation methods allow extended self life and sustained performance.

(492f) The Algebra of Sustainability: A Re-Casting of Metrics for the Artisanal Brewery
Jurado, J.,
Originally presented on: 11/6/2013 14:20:00 - 14:42:00

An infrequently tested proposition from economics is that competitive pressure will force breweries to be efficient. Key Ratios and **Sustainable development indicators** (SDI) help focus an approach to sustainability into real and quantifiable action. An attempt to create an aggregate measure of various aspects of sustainability suggests a profile of indices that provide a technical perspective that is global, like World Class Manufacturing, but with transparent numerical quantification. By excluding social factors, often a key part of Sustainability considerations (ie, 'community involvement'), calculation and presentation of SDIs is made easier and clear. Key ratios such as MJ/hL, g CO2/L or hL water/hL of finished beer help the brewery understand basic primary resource drivers of the enterprise, and if metered sufficiently, of the underlying unit operations within the enterprise. A family of key ratios comprise a new aggregate SDI, and in which specific sustainable developments at breweries can be seen to influence the SDI. The desired result is that competitive pressure in reducing SDI will motivate breweries to commit in more-sustainable investments in their production operations. Independent craft breweries have supported this development work with travel grants for the Presenter to obtain order-of-magnitude scaling data.

(492g) Economic and Environmental Sustainability in Single-Use Versus Multi-Use Equipment Decision-Making for Bioprocesses

Cheng, A. C., Carnegie Mellon University
Przybycien, T., Carnegie Mellon University
Originally presented on: 11/6/2013 14:42:00 - 15:04:00

Single-use (SU), or disposable, equipment is becoming increasingly deployed in industrial bioprocesses. The economic and environmental benefits of using SU equipment relative to multi-use (MU), or glass and stainless steel, equipment that are typically cited include the significantly lower capital costs, the smaller physical footprint of plastic versus stainless equipment, the ability to campaign SU equipment easily, the reduction in opportunities for cross-contamination and the increased throughput and reduced environmental footprint provided by the elimination of clean-in-place and sterilize-in-place cycles. While a number of economic studies and life cycle assessments have been reported for SU processes, a holistic sustainability analysis is lacking. We have begun to address this need by examining economic performance metrics (cost of goods, return on investment, payback time and net present value) and environmental performance metrics (mass indexes and environmental impact indexes) for SU versus MU processes in the face of uncertainty and for hybrid SU/MU processes. We will report on our evaluation of processes to produce high-value (monoclonal antibodies), medium-value (biosimilars) and low-value (industrial enzymes) products.

(529a) Lignocellulosic Biomass Pyrolysis Product Comparison From a Micropyrolyzer, Batch and Continuous Reactor Systems

Garrido, R., Villanova University
Hammer, N. L., USDA
Zmiewski, A., Villanova University
Satrio, J. A., Villanova University
Originally presented on: 11/6/2013 15:15:00 - 15:31:00

Lignocellulosic Biomass Pyrolysis product comparison from a micropyrolyzer, batch and continuous reactor systems

Rene Garrido, Nicole L. Hammer, Alexander Zmiewski, and Justinus A. Satrio

Department of Chemical Engineering, Villanova University

ren.garrido@villanova.edu, nhammer@villanova.edu: justinus.satrio@villanova.edu

This project aims at advancing the current state of knowledge of biomass pyrolysis by developing an understanding of the pyrolytic reaction mechanisms of different feedstocks, and to find common grounds between different lignocellulosic biomass and system configurations.

To understand these differences three successive scales of pyrolysis equipment have been implemented; Stage I, a micro-pyrolyzer, Stage II, a small-scale tubular batch reactor, and Stage III, a fluidized-bed continuous reactor. Understanding the differences on reaction mechanisms for each of the stages is a key element to properly identify and determine the products obtained from pyrolysis. To separate primary from secondary reactions, a μg scale pyroprobe system combined with gas chromatography and mass spectrometry (Py-GCMS) is used to study the chemical composition of volatile products from previously hydrolyzed and torrefied biomass. Effects of the severity of hydrolysis pretreatment and torrefaction on the chemical composition and yield distribution of the fast pyrolysis products will be evaluated. The chemical composition study of pyrolyzed biomass would then aid in the understanding of the chemistry behind the different compounds to further enhance the process of upgrading bio-crude oil.

For a batch reactor, a tubular furnace is used to pyrolyze the samples, and the products will be compared with a continuous fluidizing bed reactor. Although the reaction mechanisms are different, and the biomass particles in the fluidization medium have greatly enhanced heat and mass transfer in fluidized beds reactors. This is the first step to understand and find common ground between the different feedstocks, and to move from lab scale to a pilot scale reactor.

This study investigates common Reed Grass *(Phragmites Australis)*, and Pinewood as examples of different sources of lignocellulosic biomass as potential feedstock for bio-oil production.

Phragmites represents grassy biomass; it is a prevalent waste on Villanova's campus and around the world. It is potentially a valuable energy and chemical feedstock due to its high yield (18-28 tons of biomass per acre per year). Pinewood represents hardwood biomass with high cellulose content and for our study it is the control system in which all other biomasses will be related to.

Based on the results obtained by analytical pyrolysis this study will focus on obtaining bio-crude oil from *Phragmites australis* and Pinewood using a batch reactor and compare the results from a fluidized bed continuous reactor. Through this study, groundwork is established as a preliminary attempt to understand how various feedstocks can be used for biomass conversion to more useful energy sources.

(529b) Spatio-Temporal Carbohydrate Characterization of Wood Particles During Fast Pyrolysis

Paulsen, A. D., University of Massachusetts Amherst
Hough, B.,
Schartz, D.,
Pfaendtner, J., University of Washington
Dauenhauer, P. J., University of Massachusetts Amherst
Originally presented on: 11/6/2013 15:31:00 - 15:47:00

Fast pyrolysis provides an inexpensive chemical process for converting lignocellulosic biomass including wood fibers and grasses to a liquid intermediate called bio-oil which can be transported and upgraded to chemicals and fuels. The complexity of fast pyrolysis arises from the feedstock which is comprised of several biopolymers (cellulose, hemicellulose, and lignin) integrated within a cellular microstructure. During rapid heating, cellulose and biopolymers rapidly degrade to a mixture of short-chain oligomers and eventually volatile components and aerosols which are collected as bio-oil [1,2,3]. When wood fibers are heated in pyrolysis, heat and mass transfer limitations within the cellular lignocellulosic structure produce dramatic temperature gradients, which ultimately dictate the local reaction conditions and the distribution of pyrolysis products. The complexity of characterizing the reacting particle chemistry and transport phenomena requires a large number of kinetic parameters and has been identified as one of the major fundamental challenges in research of fast pyrolysis of biomass [4].

In this work, we aim to improve reaction-transport pyrolysis models of reacting woody biomass by demonstrating a new experimental technique that characterizes the composition of woody biomass as it pyrolyzes to bio-oil. Using diffuse reflectance spectroscopy, the carbohydrate composition of Yellow Poplar (*Liriodendron tulipifera*) is characterized in position and time through a particle exhibiting heat transfer dominant in only one spatial dimension. Compositional profiles exhibit a propagating reaction front consistent with heat-transfer limited (high Bi) pyrolysis conditions. A range of reaction conditions are examined, and the resulting data are described using a one-dimensional reaction-transport model using existing lumped-kinetic reaction chemistry. Sensitivity analysis of the resulting model and comparison of the prediction with experimental spatio-temporal compositional data allows for robust characterization of wood fiber pyrolysis.

References

[1] M.S. Mettler, D.G. Vlachos, P.J. Dauenhauer, *Energy & Environmental Science* **2012**, 5, 7797-7809.

[2] V. Agarwal, P.J. Dauenhauer, G.W. Huber, S.M. Auerbach, *Journal of the American Chemical Society* **2012**, 134(36), 14958-14972.

[3] A.R. Teixeira, K.G. Mooney, J.S. Kruger, C.L. Williams, W.J. Suszynski, L.D. Schmidt, D.P. Schmidt, P.J. Dauenhauer, *Energy & Environmental Science* **2011**, 4, 4306.

[4] M.S. Mettler, D.G. Vlachos, P.J. Dauenhauer, *Energy & Environmental Science* **2012**, 5, 7797-7809.

(529c) Numerical Investigation of Multiphase Dynamic Effects in Catalytic Upgrading of Biomass Pyrolysis Vapor

Capecelatro, J. S., Cornell University
Desjardins, O., Cornell University
Pepiot, P., Cornell University
Jarvis, M., National Renewable Energy Laboratory
Foust, T. D., National Renewable Energy Laboratory
Originally presented on: 11/6/2013 15:47:00 - 16:03:00

A recurring challenge among the variety of existing biomass-to-biofuel conversion technologies is the need to ensure optimal and homogeneous contact between the various phases involved. One example can be found in the catalytic upgrading of bio-oil vapors, a bi-phasic process critical in pyrolysis-based conversion, typically performed in circulating fluidized bed (CFB) reactors. While the upgrading process is expected to strongly depend on the contact time between raw gas and catalysts, and on catalysts deactivation over time, the formulation of robust design rules from an empirical standpoint alone remains difficult, due to the wide range of granular flow regimes coexisting within the reactor, from near close-packing at the bottom of the reactor to more dilute conditions in the riser. The issue is amplified by the highly unsteady and turbulent nature of gas-solid flows. In particular, clusters of particles are known to form in risers, which can significantly decrease the overall catalytic activity in the reactor. In this work, a Lagrangian particle tracking tool recently developed in the large eddy simulation (LES) framework of NGA is used to quantify the role of cluster dynamics on biomass pyrolysis vapor catalytic upgrading in risers. This numerical framework has been extensively validated for a range of particle-laden flows, in particular, cluster fall velocities in risers of CFB reactors have been shown to match experimental correlations (Fig.1). Two configurations are investigated and compared: an homogeneous system for which the catalyst-pyrolysis vapor contact time is statistically uniform, and a pseudo-two-dimensional riser, such as shown in Fig.2, combining all granular flow regimes of interest.

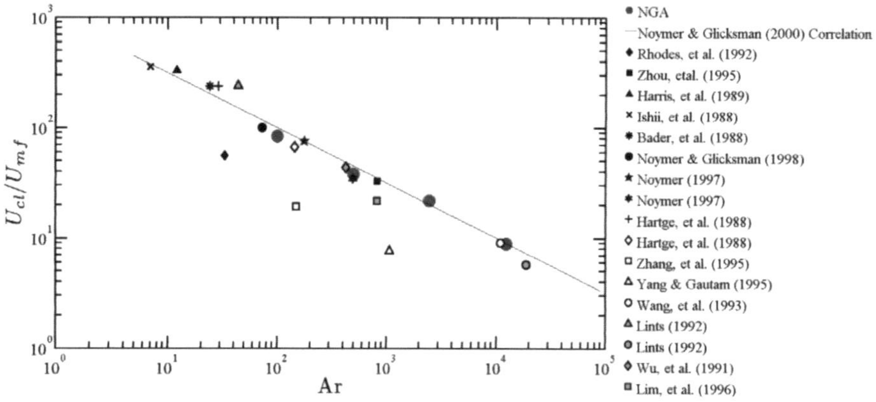

Figure 1: Mean cluster fall velocity normalized by the minimum fluidization velocity as a function of Archimedes number, a non-dimensional number that characterizes the flow inertia due to density differences between the gas and solid phases. The solid line indicates a correlation developed from a range of experimental studies by Noymer & Glicksman (2000).

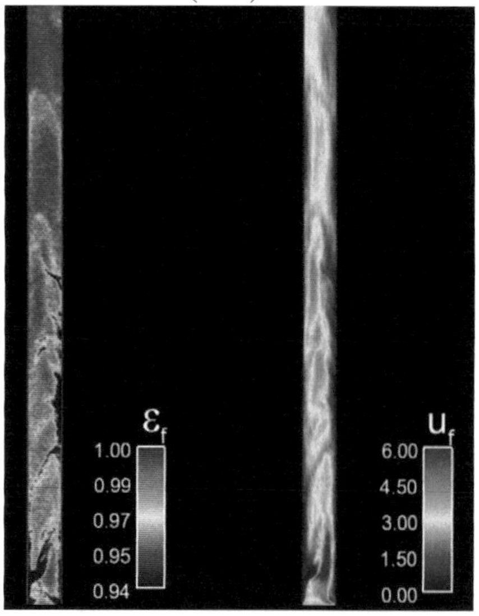

Figure 2: Pseudo two-dimensional simulation of a riser showing gas volume fraction (left) and gas velocity magnitude (right). Particles are injected at the bottom where the gas inlet is located.

(529d) Catalytic Effects of MgSO4 and ZnCl2 On Fast Pyrolysis of Pine for the Production of Bio-Oils

Yuan, L.,
Liu, Y. Q., Chevron Technology Ventures LLC
Lv, D. C., Xiamen University
Wang, D., Xiamen University
Originally presented on: 11/6/2013 16:03:00 - 16:19:00

The effects of $MgSO_4$ and $ZnCl_2$ as catalysts on fast pyrolysis of pine were investigated in a bench-scale bubbling fluidized bed reactor. The study showed that both $MgSO_4$ and $ZnCl_2$ could shift the pyrolysis of pine to a lower temperature zone, that is, most of pyrolytic reactions for cellulose and hemicellulose would happen at lower temperature while lignin at higher temperature, which results in different composition of bio-oils from that without adding any of these two salts. Compared to the bio-oils obtained without catalysts (base case), glucopyranose in bio-oils was increased to 20.69 % from less than 1% (base case) when 20 wt% of $MgSO_4$ was added; while ketones decreased from 26.71 % (base case) to 8.41 %, 2-Propanone, 1-hydroxy from 9.94 % (base case) to 0.62 %. When 30 wt% of $ZnCl_2$ is added, glucopyranose in bio-oils hiked to 20.21 % from less than 1% (base case); while phenols reduced to 6.25% from 27.09 % (base case). The pH of bio-oils rises whether $MgSO_4$ or $ZnCl_2$ was added. This study suggests that by adding specific salts into biomass, the composition of bio-oils obtained from fast pyrolysis can be of greatly different. Therefore, we can increase the content of certain chemicals in bio-oils on purpose, which could make the extraction of value-added chemicals from bio-oils more feasible and economical.

(529e) Pyrolysis of Alkali Pre-Treated Biomass and Biomass Constituents

Case, P., University of Maine
Stemmler, E., Bowdoin College
DeSisto, W. J., University of Maine
Wheeler, M. C., University of maine
Originally presented on: 11/6/2013 16:19:00 - 16:35:00

Although upgrading of pyrolysis oil has been extensively studied, it is still plagued with low product yields and significant catalyst deactivation. These issues are largely because of the instability of oxygenated species in pyrolysis oil, which leads to considerable char formation by polymerization at the elevated temperatures that are required for hydrodeoxygenation. In the current study, we eliminate compounds that cause polymerization by pre-treating sawdust with various alkali compounds before pyrolyzing it in a fluidized bed reactor. The resulting stabilized oils have significantly lower oxygen content than conventional pyrolysis oil. The higher heating values of the oils are in the range of 40 MJ/kg, and the energy yields exceed those commonly achieved in catalytic fast pyrolysis. Detailed analysis of conventional and pre-treated pyrolysis oils is presented and potential reaction mechanisms are discussed.

(529f) Study On the Aging Mechanism of Pyrolysis Bio-Oil

Meng, J., North Carolina State University
Tilotta, D., North Carolina State University
Adhikari, S., Auburn University
Kelley, S., North Carolina State University
Park, S., North Carolina state university
Originally presented on: 11/6/2013 16:35:00 - 16:51:00

Session 16004 Biomass Pyrolysis I

AIChE 2013

November 3 ~ 8, 2013

San Francisco, CA

Study on the Aging Mechanism of Pyrolysis Bio-oil

Jiajia Meng*,[1] David Tilotta,[1] Sushil Adhikari,[2] Stephen Kelley,[1] and Sunkyu Park[1]

[1] Department of Forest Biomaterials, North Carolina State University, Raleigh, NC 27695

[2] Biosystems Engineering Department, Auburn University, Auburn, AL 36849

*Presenting author: jmeng@ncsu.edu

The unstable nature of pyrolysis oil produced from biomass fast pyrolysis is one of the key issues that limit its application and upgrading, as aged bio-oil typically shows increased water content, viscosity, and phase separation. The aging phenomena are considered due to the polymerization of bio-oil components under acidic and thermal conditions catalyzed by the mineral compounds contained in char particles in the bio-oil. Current research on bio-oil aging is focused on characterizing the behavior together with seeking methods to slow down the aging speed. However, little research was reported on investigating the molecular-level polymerization mechanism of pyrolysis bio-oil during the aging. In this study, accelerated aging of the bio-oil produced from loblolly pine was performed with different aging time and temperature and with or without additives to access the bio-oil thermal stability by measuring water content, viscosity, acidity, molecular weight distribution, and chemical composition. With the chemical analysis of aged bio-oils, it was found that ether soluble fraction of the bio-oil was significantly reduced during the aging process and glycolaldehyde is the most reactive compound in that fraction. The pyrolytic lignin fraction, especially high molecular weight lignin, was found to gradually increase, which suggests the reaction between the compounds of ether soluble fraction and the lignin fraction. Because of the great reduction of glycolaldehyde concentration during the aging, synthetic bio-oil with different concentration of glycolaldehyde was mixed with pyrolytic lignin and aged at 80 °C to further study the role of glycolaldehyde on bio-oil aging chemistry. To elucidate the polymerization chemistry between ether soluble fraction and pyrolytic lignin fraction, model compounds, such as ^{13}C labeled glycolaldehyde and ^{13}C labeled phenolic compounds (phenol and guaiacol type), were reacted in a similar condition to that of accelerated aging. In addition, other suggested mechanisms of bio-oil aging will be reviewed and discussed in detail.

(529g) Effects of Post-Condensation Centrifugal Filtration On Pyrolysis Oil

Varadarajan, A., Mississippi State University
Walters, K. B., Mississippi State University
Originally presented on: 11/6/2013 16:51:00 - 17:07:00

The effects of post-condensation filtration of the crude pyrolysis oil with relatively high water content and solids content was studied. Tests were performed using a pilot plant-scale centrifuge filtration system. The centrifuge can be used as a three-phase separator [light liquid-heavy liquid-solids] or a two-phase clarifier [liquid-solid]. The testing was performed by varying the parameters such as separation time and displacement time which affects the separation efficiency of the centrifuge. In addition, the impact of increased temperature prior to and during the filtration process was examined. Effect of feed temperature on the separation efficiency of the centrifuge and the consequent physical and chemical property change on the filtered bio oil was performed. The filtered samples were collected and characterized. To determine the water content, the Karl Fischer titration method was utilized. Fourier transform infrared (FTIR) spectroscopy was used to determine chemical composition. Gel permeation chromatography was used to determine molecular weight distribution. Particle size distribution in the filtered samples was found using optical microscopy.Parameters were examined for the centrifugal filtration of post-condensation pyrolysis oil to optimize the particulate loading, processing time, and physicochemical stability for downstream processing, including upgrading

(529h) Transportable Fast-Pyrolysis Process for Distributed Conversion of Waste Biomass to Renewable Liquid Fuels

Yelvington, P. E., Mainstream Engineering Corporation
Schwartz, N., Mainstream Engineering Corporation
Zastrow, D., Mainstream Engineering Corporation
Amundsen, T., Mainstream Engineering Corporation
Originally presented on: 11/6/2013 17:07:00 - 17:23:00

In Florida alone, sugarcane bagasse, orange residue from juicing, and logging/mill residues are produced at a rate exceeding 16,000 dry tons per day. However, these agricultural residues are often spread over large areas and are difficult to access, particularly the logging residues. Rather than trucking large quantities of low-density biomass to a massive centralized biofuel production plant, the distributed conversion of agricultural residues into bio-oil is an attractive alternative for sustainable biofuel production. Opportunities also exist for the conversion of mixed waste to biofuel at forward-deployed military outposts, where the true delivered cost of fuel often exceeds $15 per gallon. Mainstream Engineering (MEC) recently commissioned a highly instrumented fast-pyrolysis pilot plant (1 ton/day) to demonstrate small-scale conversion of waste organic material to bio-oil. The pilot reactor was designed for flexibility and tight process control, as we explore the product yield and quality for a variety of feedstocks. Testing to date has used pine sawdust or a granulated mixed waste surrogate consisting of corrugated cardboard, office paper, and several plastics.

MEC is working to commercialize the portable process at a 10 ton/day scale for commercial applications and at the <1 ton/day scale for military applications. Our efforts have focused on improving the reactor yield by minimizing secondary cracking and improving the energetics with tight heat integration. Both improvements are geared towards achieving viable process economics at a small thoughput. We have also experimentally investigated the pre-processing (size reduction and drying) and end-use utilization of raw bio-oil in a burner/boiler. Results will be presented on our pilot reactor testing, techno-economic analysis of the portable reactor approach, and initial results from our efforts to mildly upgrade raw bio-oil to a renewable residual-grade fuel oil.

(529i) Biomasspyrolysisrefinery: Product Stability of Intermediates and Products

Schwaiger, N., Graz University of Technology
Feiner, R., Graz University of Technology
Pucher, H., Graz University of Technology
Steiner, L., Graz University of Technology
Ellmaier, L., BDI-Bioenergy International AG
Pucher, P., BDI BioEnergy International AG
Siebenhofer, M., Graz University of Technology
Originally presented on: 11/6/2013 17:23:00 - 17:39:00

The BiomassPyrolysisRefinery is a dual step concept for biomass liquefaction and hydrogenation, intending production of liquid energy carriers. First step of this concept is liquid phase pyrolysis (LPP). During pyrolysis lignocellulose is dehydrated to form biochar and pyrolysis oil (Schwaiger et al., 2011, 2012) The LPP process has been scaled up to a capacity of 100 kg/h biomass (BDI bioCRACK facility) at OMV refinery Schwechat. The second step of the BiomassPyrolysisRefinery is product upgrading. Pyrolysis oil can pass a dewatering step to raise the calorific value. It may also be deoxygenated and hydrogenated. Biochar is separated from the heat carrier after liquid phase pyrolysis and then hydrogenated.

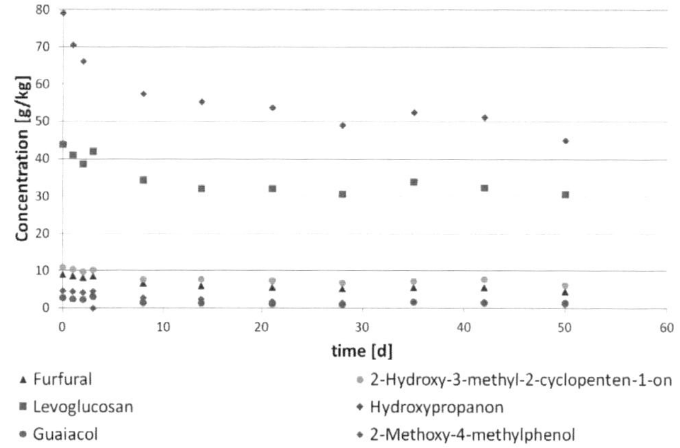

Figure 1: Stability of selected substances from liquid phase pyrolysis during storage at ambient temperature

During liquefaction biogenous material passes several heating and storing steps. Therefore sufficient chemical stability of intermediates and products is needed. Stability of pyrolysis oil for storing and upgrading purposes may be very limited (Fahmi et al., 2008; Kim et al., 2012), because of polymerization. The molar mass, viscosity and water content may rise significantly. The storing properties of liquid phase pyrolysis oil, upgraded liquid phase pyrolysis oil and liquified biochar at ambient temperature was monitored by GC-MS, GC-TCD, SEC and pH, and TAN analysis. Density and viscosity were recorded.

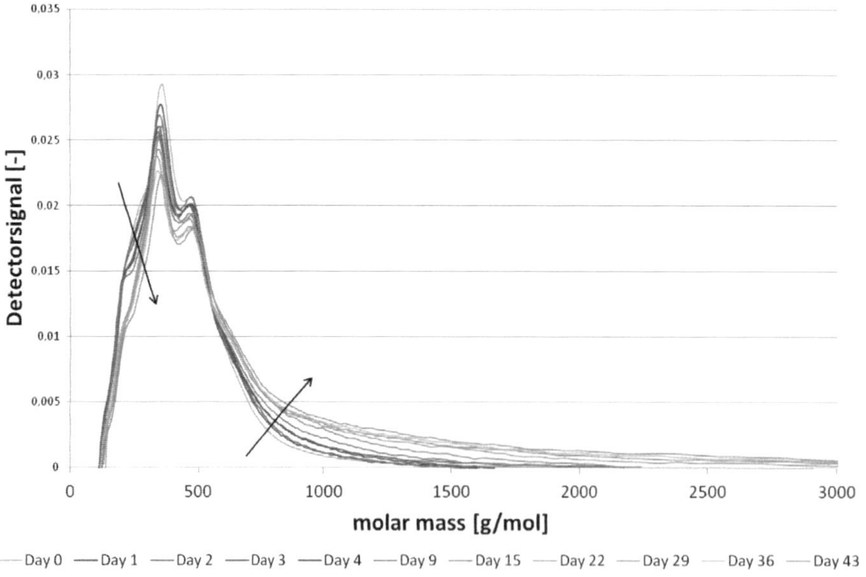

Figure 2: Liquid phase pyrolysis oil; molar mass versus storage time

Figure 1 shows the concentration of liquid key substances from liquid phase pyrolysis. During the first week of storage nearly one third of highly oxygenated substances vanish, afterwards degradation may cease. The same trend of property changes in the first days after pyrolysis is recorded with SEC-analysis. Figure 2 shows stabilization of molecular mass within day 15.

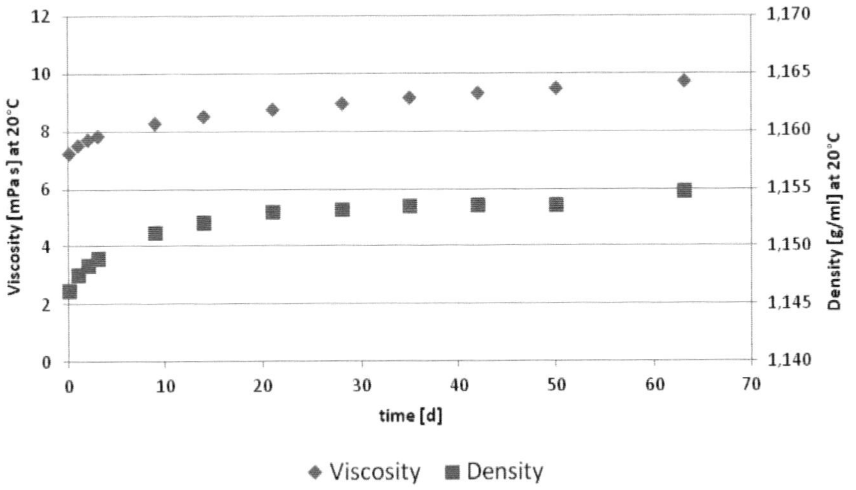

Figure 3: Viscosity and density at T = 20°C of liquid phase pyrolysis oil

Viscosity and density trends, shown in figure 3, confirm the findings. Liquid phase pyrolysis oil undergoes ageing through repolymerization within the first 15 days of storage. After upgrading stability of pyrolysis oil and liquified biochar is significantly better. Compared with flash pyrolysis oil the stability of liquid phase pyrolysis oil is seemingly achieved through separation from char based particulate matter and inorganics.

Fahmi, R., Bridgwater, A., Donnison, I., 2008. The effect of lignin and inorganic species in biomass on pyrolysis oil yields, quality and stability. Fuel 87, 1230--1240.

Kim, T.-S., Kim, J.-Y., Kim, K.-H., Lee, S., Choi, D., Choi, I.-G., Choi, J.W., 2012. The effect of storage duration on bio-oil properties. Journal of Analytical and Applied Pyrolysis 95, 118--125.

Schwaiger, N., Feiner, R., Zahel, K., Pieber, A., Witek, V., Pucher, P., Ahn, E., Wilhelm, P., Chernev, B., SchrÃ¶ttner, H., Siebenhofer, M., 2011. Liquid and Solid Products from Liquid-Phase Pyrolysis of Softwood. BioEnergy Research 4, 294--302.

Schwaiger, N., Witek, V., Feiner, R., Pucher, H., Zahel, K., Pieber, a, Pucher, P., Ahn, E., Chernev, B., Schroettner, H., Wilhelm, P., Siebenhofer, M., 2012. Formation of liquid and solid products from liquid phase pyrolysis. Bioresource technology 124, 90--4.

(529j) Biomass Gasification: Effect of Sulfur Compounds On Catalytic Tar Removal Activity Using Nickel-Clay Catalysts

Vutukuri, J., University of Louisiana at Lafayette
Buchireddy, P., University of Louisiana at Lafayette
Bricka, R. M., Mississippi State University
Guillory, J., University of Louisiana at Lafayette
Bajpai, R., University of Louisiana at Lafayette
Zappi, M. E., University of Louisiana at Lafayette
Originally presented on: 11/6/2013 17:39:00 - 17:55:00

Biomass gasification is one of the thermochemical routes of converting biomass to fuel gas. This fuel gas can be used in various possible applications such as generation of heat or electricity, synthesis of liquid transportation fuels, production of hydrogen, synthesis of chemicals and generation of electricity in fuel cells. However, this fuel gas needs to be cleaned from the impurities like particulates, ammonia, hydrogen sulfide, hydrogen chloride, alkali metals, metals, and tars. The presence of tars in the fuel gas is one of the main technical barriers in the biomass gasification development. Tar buildup cause cracking in the pores of filters, forming coke and plugging the filters, and condensing in the cold spots and plugging the lines, resulting in serious operational interruptions.

Nickel (Ni) supported clays have been developed and evaluated for its activity towards tar removal at Mississippi State University. The results from these studies suggested that Ni-Clay catalysts have very good activity for tar removal. However, the influence of sulfur compounds, although, present in very low concentrations has not been evaluated. Hence, this study focuses on the effect of sulfur compounds on the catalyst performance for removal of tars from syngas. Literature suggests that the presence of sulfur in the fuel gas decreases the catalytic activity of the Ni catalysts due to strong chemisorption of sulfur on the Ni surface forming Nickel Sulfide, which is inactive. Although, the presence of sulfur has shown to have an inhibiting effect on the catalytic activity, the effect of sulfur on tar removal using Ni-Clays has not been investigated till date. This study investigates the influence of sulfur compounds on tar reforming activity of Ni Clay catalysts. A plug flow reactor has been designed and fabricated to perform these studies. Experiments are being conducted using naphthalene as a tar model compound and hydrogen sulfide as a sulfur compound in syngas. The effect of hydrogen sulfide partial pressure, temperature(650-850° C), steam to carbon ratio, and residence time on naphthalene removal activity is being evaluated, the results of which will be presented. Also, the influence of catalyst support, clay in this study is being investigated and will be compared with commercial silica/alumina supports.

(546a) Controllability of Complex Networks for Sustainable System Dynamics

Benavides, P. T., University of Illinois at Chicago
Diwekar, U. D., Vishwamitra Research Institute
Cabezas, H., U.S. Environmental Protection Agency
Originally presented on: 11/6/2013 15:15:00 - 15:37:00

Successful implementation of sustainability ideas in ecosystem management requires a basic understanding of the often non-linear and non-intuitive relationships among different dimensions of sustainability, particularly the system-wide implications of human actions. This basic understanding further includes a sense of the time scale of possible future events and the limits of what is and is not likely to be possible. With this understanding, systematic approaches based on control theory can then be used to develop policy guidelines for the system. Therefore, controllability of the system is very important to determining long term sustainability of the system. A recent article in the journal Nature (1) presents a new analytical approach to study the controllability of complex systems. We apply this approach to three dynamic systems developed to study the sustainability of our planet. These three systems consist of an ecosystem based on wild and domesticated compartments, the ecosystem model with an industrial system and a simple economic model, and an integrated model involving the ecosystem, industrial systems, and energy producers. We argue that controllability of this system is linked with the long term sustainability, and we present our arguments in the light of previous studies of these systems.

(1) Liu, Y. Y., Slotine, J. J., Barabási, A. Controllability of complex networks. *Nature.* **2011**, *473*, 167

(546b) Sustainable Process Decision Making With Greenscope

Smith, R. L., U.S. Environmental Protection Agency
Ruiz-Mercado, G. J., U.S. Environmental Protection Agency
Gonzalez, M. A., U.S. Environmental Protection Agency
Originally presented on: 11/6/2013 15:37:00 - 15:59:00

The U.S. Environmental Protection Agency is developing a method and tool called GREENSCOPE (Gauging Reaction Effectiveness for the Environmental Sustainability of Chemistries with a multi-Objective Process Evaluator) to support sustainable processes. GREENSCOPE can be used to evaluate existing processes or in the design of new ones. With 140 indicators for objectives in four basis areas: environment, economics, energy, and efficiency, GREENSCOPE provides information on process sustainability. Results for the various objectives show where processes are performing well and where improvements could be made.

One of the most difficult issues of dealing with multi-objective results is determining a scientific way to incorporate the process information and user preferences. These preferences represent the values of the decision maker, and often scientists and engineers are uncomfortable with the methods available for integrating values into objective and quantifiable science. While the existence of values in decision making cannot be eliminated, one can navigate a path that provides scientific clarity.

Much of the effort needed to provide clarity in decision making has already been incorporated into the GREENSCOPE methodology. Previous work has developed absolute sustainability limits for the indicators in GREENSCOPE [1], which provides ranges that delineate worst case and target values. Decision makers can define marginal rates of substitution (i.e., points of equal tradeoffs) between the range-bound indicators. Inputs such as the marginal rates of substitution and GREENSCOPE indicator results provide equations that describe decision making for multi-objective process evaluations, with a strictly mathematical method solving the equations for various inputs [2]. The method will be exemplified through various results for a terephthalic acid process.

[1] Ruiz-Mercado, G. J.; Smith, R. L.; Gonzalez, M. A., Sustainability Indicators for Chemical Processes: I. Taxonomy. *Ind. & Eng. Chem. Res.,* 51, 2309-2328 (2012).

[2] Smith, R.L.; Ruiz-Mercado, G.J., A Method for Decision Making using Sustainability Indicators. Submitted, *Clean Tech. & Env. Policy* (2012).

(546c) Innovative Chemical Engineering Applications to Control Air Emissions From Oil Refineries

Nair, D. C., Belco Technologies Corporation

Originally presented on: 11/6/2013 15:59:00 - 16:21:00

Wet gas scrubbing technology is a well proven and reliable method for controlling flue gas emissions of particulate, SOx and NOx to the atmosphere from fluid catalytic cracking units (FCCUs) used in oil refineries.

This paper will show that there has been a viable technology for the past several years that can minimize environmental impacts using innovative chemical engineering practices along with design of unique proprietary equipment that provides non-stop operation for at least five years. The paper will provide details of the unique design that have been used at well over 50 installations at oil refineries in the U.S and overseas. A description of the wet scrubbing system utilized for this project is described. Specifically, this description will address the methodology for removing particulate, SO_2 and NOx and the process for treating the liquid effluent from the wet scrubbing system. Also addressed are the aspects of three designs that allow for achieving different levels of particulate emissions.

This paper is also an attempt to compare a presentation made at the AIChE Spring meeting that a Genetic Algorithm Approach for Optimal Control of Air Emissions from an Oil Refinery was possible.

The reduction of environmental impacts is an essential step towards the improvement of sustainability for any given producer.

(546d) Direct and Indirect Emissions From Mobil Air Conditioner Units in Transportation By Truck and Rail

Blowers, P., The University of Arizona
Dougless, A., U of Arizona
Gerwe, B., U of Arizona
Mesta, A., U of Arizona
Rugel, C., U of Arizona
Originally presented on: 11/6/2013 16:21:00 - 16:43:00

The U.S. EPA has announced repeatedly that hydrofluorocarbons (HFCs) will become regulated, not for their direct impacts on the environment, but because of their contributions to global climate change. HFCs are used in many industrial processes as solvents and refrigerants because they are relatively inert, inexpensive and have physical properties that allow them to function well. The use of HFCs is largest in the refrigeration sector. R-134a (1,1,1,2-tetrafluoroethane is the dominant refrigerant used currently in mobile air conditioner systems and has emissions of about 38.6 x 10^6 kg/year per year from all sources, with approximately 47% of that being due to vehicle emissions. The European Union took steps in January 2011 to eliminate fluorinated greenhouse gases in certain classes of vehicles with 100 year GWP's greater than 150 as defined by the third assessment report adopted by the IPCC, including R-134a. Currently passed legislation in Europe also calls for a complete phase-out by January 2017.

Mobile air conditioning units (MAC) are in use in several applications, with their use in passenger cars having been evaluated for greenhouse gas emissions in the prior literature. If HFCs are banned as a class of compounds, alternatives will need to be developed or exploited. While MAC and environmental impacts have been investigated for passenger cars, there are other MAC units where the temperature constraints and demands may lead to a more restricted range of possible refrigerants. Mobile air conditioning units used in the shipping of temperature sensitive products, like foods and medication, is common on both tractor trailers and on refrigeration, reefer, units installed on rail cars. With the large number of US and worldwide refrigerated shipping units, evaluations of the emissions for this application of refrigeration need to be obtained before deciding if banning hydrofluorocarbons is indeed better for the environment.

This work uses chemical engineering heuristics and design parameters, combined with actual operating conditions and design specifications of manufactures to estimate the direct and indirect greenhouse gas emissions for MAC using diesel powered reefer units in transport. We find that many refrigerants that can be used in dedicated non-mobile units where water cooling is possible fail for this application. The remaining refrigerants are small in number and it appears that banning HFCs would be unattractive due to the indirect emissions.

(546e) Multi-Objective Optimization of International Economies Via Multi-Regional Input-Output Analysis: Application to the US Economy

Pascual, J.,
Guillen, G., Universitat Politecnica de Catalunya
Jimenez, L., University Rovira i Virgili
Cortes, D., University Rovira i Virgili
Originally presented on: 11/6/2013 16:43:00 - 17:05:00

Multi-objective optimization of international economies via multi-regional input-output analysis: Application to the US economy

J. Pascual-Gonzalez[a], *G. Guillen-Gosalbez*[a][*] — *and L. Jimenez*[a]

[a]Departament d'Enginyeria Quimica, Universitat Rovira i Virgili, Av. Paisos Catalans 26, Tarragona 43007, Spain

Abstract

In recent years, new environmental policies have been developed to reduce the environmental impact of human activities. Designing these policies in an effective manner is challenging because it requires a detailed knowledge of the whole economic and environmental system and the way in which they interact with each other.

In terms of economics, input-output analysis (IO) is widely used to predict changes in an entire economy according to changes in the demand of a single (or several) sectors. This method studies the effects that different sectors have on the economy as a whole. Purely economic IO tables can be modified so as to include environmental aspects, giving rise to environmentally extended input-output tables (EEIO). These tables contain information on the environmental assessment of the economic transactions taking place in a region, but provide no guidelines on how to reduce such an impact.

We present in this work a systematic approach to reduce the environmental impact of a whole economy. Our strategy combines multi-objective optimization and multi-regional input-output models within a single unified framework that allows identifying the activities whose control and regulation can lead to larger reductions of impact at a minimum decrease in gross domestic product (GDP).

We have applied our tool to the economy of the United States in order to identify the best policies to be implemented in practice for optimizing simultaneously the economic and environmental performance. We have employed in the calculations the World Input-Output Database (WIOD), which was originally developed to analyze the effects of globalization on trade patterns, environmental pressures, and socio-economic development across a wide set of countries. WIOD describes the economic inputs and outputs in monetary terms of 35 manufacturing sectors covering 27 EU countries and 13 other major countries in the world for the period 1995 to 2009.

Our approach provides valuable insight into how to reduce the impact globally. By adopting this birds' eye view of a whole economy, it is possible to determine how the final demand (and the total production) of every sector in an economy should be modified so as to improve simultaneously the economic and environmental performance of an economy.

---

[*] Corresponding author: tel. +34-977 55 86 18 email: gonzalo.guillen@urv.cat

(546f) Urban Systems and Sustainability: Comparative Analysis of Multiple Metropolitan Statistical Areas in Ohio, USA

Gonzalez-Mejia, A., US EPA - ORISE
Eason, T., U.S. Environmental Protection Agency
Cabezas, H., U.S. Environmental Protection Agency
Suidan, M., University of Cincinnati
Originally presented on: 11/6/2013 17:05:00 - 17:27:00

Urban systems have a number of factors (e.g., economic, social and environmental) that can potentially impact their growth, change and transition. As such, assessing and managing these systems is a complex challenge. While, tracking trends of key variables may provide some insight, identifying the critical characteristics that truly impact the dynamic behavior of these systems is difficult. As an integrated approach to evaluate real urban systems, this work contributes to the research on scientific techniques for assessing sustainability. Specifically, it proposes a practical methodology based on the estimation of dynamic order, for identifying stable and unstable periods of sustainable or unsustainable trends using Fisher Information (FI). In this work, the dynamic behavior of main six Metropolitan Statistical Areas (MSAs) in Ohio was evaluated by using 29 social and 11 economic variables to characterize each system from 1970 to 2009. For a complete regional sustainability analysis, these Ohio MSAs were compared to each other according their stability and population size over the period of study, to assess the impact of social and economic characteristics on system stability at geographic scales, which are critical to policy and management. Results from this multi-year assessment indicated that the Columbus MSA is the most stable of the MSAs understudy, followed by the Cincinnati MSA. Further, while there were unique trends in each MSA, changes in family structure and income appeared to correspond with the patterns experienced, particularly near the end of the study period (2000 – 2009). While the approach presented is not a cause and effect analysis, it points out characteristic features of complex systems that are otherwise very difficult to see by inspecting the individual variables, which characterize the system over time. Further, this approach intends to aid in locating significant deviations in the system's trajectory and determine whether these shifts corresponds to a new path characterized by sustainable (e.g., desirable growth rate) or unsustainable trends (e.g., decrease in the income growth rate). This work contributes holistically to the research on evaluating complex urban systems and affords the ability to identify and distinguish local and regional drivers of change. The methodology described can be used by scientists, planners and policy makers interested in evaluating dynamic changes in complex systems and identifying potential drivers at multiple spatial and temporal scales.

(546g) Environmental Impact Minimization of Vaccine Supply Chain for Developing World

Mousavi, B., University of Colorado Denver
Karunanithi, A. T., University of Colorado Denver
Originally presented on: 11/6/2013 17:27:00 - 17:49:00

People in low and middle income countries are most vulnerable to environmental, climate change and human health impacts. Vaccines offer enormous public health benefits (avoid disease and reduce treatment costs) while emissions from vaccine supply system result in environmental/health impacts (Green house gases and criteria pollutants) in particular, immunization supply systems of low and middle-income countries are not optimized to reduce environmental impacts. Current immunization supply systems are optimized to minimize direct costs. Such an approach usually results in higher levels of pollutants leading to lower public health outcomes which in terms results in higher treatment costs to the society. This nexus is not fully understood or recognized by public health programs. A truly optimized immunization supply system will not only consider direct costs but also indirect treatment costs averted by minimizing public health burdens to the society.We will present an approach that utilizes LCA methodologies to optimize the life cycle of vaccines that include production, packaging, transport, cold storage, use and disposal in the context of developing world.

(564a) Strategies in the Sustainable Design of Active Pharmaceutical Ingredients for Solvent Recovery Systems

Savelski, M. J., Rowan University
Slater, C. S., Rowan University
Cavanagh, E. J., Rowan University
Originally presented on: 11/6/2013 15:15:00 - 15:40:00

The Chemical Engineering Department at Rowan University has partnered with several pharmaceutical companies and the US EPA to explore more sustainable pathways in the manufacturing of active ingredients and to educate future generations of engineers and decision makers in the importance of green engineering in pharmaceutical syntheses. The conducted research has evaluated greener designs for solvent recovery targeting not only the reduction of the overall solvent consumption, but the associated waste, and life cycle emissions from cradle to gate. These studies have explored greener separation processes and greener solvents using both LCA and economic metrics. Software has been developed to guide decision making and optimize processes based on these desired outcomes. When integrated correctly, solvent recovery is a viable option to reduce significantly the carbon footprint of drug synthesis as well as operating costs. Life cycle analysis of pharmaceutical manufacturing routes involving solvents such as acetonitrile, isopropanol, methanol, tetrahydrofuran, and methyltetrahydrofuran have been conducted. This paper will present the approaches taken to determine the optimal application of greener technologies to minimize emissions and make pharmaceutical manufacturing more sustainable.

(564b) Towards a Complete Solution of Allocation in Life Cycle Inventories

Hanes, R., The Ohio State University
Cruze, N.,
Goel, P. K., The Ohio State University
Bakshi, B. R., The Ohio State University
Originally presented on: 11/6/2013 15:40:00 - 16:05:00

The need for allocation arises in life cycle inventories when a process results in multiple products. In such a situation, resources and emissions embodied in the inputs need to be partitioned between the products. Popular approaches for accomplishing this task involve partitioning the embodied resources in proportion to characteristics of the outputs such as mass, energy, exergy or cost. Given the subjectiveness of this decision, many practitioners apply multiple allocation methods. Recent efforts have proposed the use of regression methods such as ordinary least squares and total least squares to solve the allocation problem. Allocation may be avoided by expanding the system boundary, but this approach can be tedious due to the need for more data.

We show that allocation can be understood as an ill-posed problem, analogous to the situation where the number of variables in simultaneous equations is larger than the number of independent equations. In such a situation, the number of solutions is infinite, and without additional information or assumptions, a single "correct" solution cannot be determined. Using a matrix framework for life cycle inventories, we devise a method of calculation whereby results from all possible allocation methods can be examined.

We use this insight to evaluate the special case where LCA is used to compare products that have the same use but are produced by different processes. Examples of such a case include comparing ethanol from various sources, gasoline, and other transportation fuels. We develop an approach to determine the sensitivity of the comparative LCA to allocation weights and show that sometimes the better product may be independent of the allocation approach. For cases when this does not happen, we propose an optimization approach to determine the "tipping curve" or the boundary across which the product preference changes.

The practical usefulness of this approach is demonstrated with two case studies. The first study compares the LCIs of 1,3-propanediol (PDO) produced from fossil fuels and from biomass. This comparison was found to be robust to allocation for nine resources and emissions out of the eleven considered: biomass PDO was preferred over fossil PDO for all choices of allocation weights. The second case study compared ethanol produced from corn stover hydrolysis, corn stover gasification, and corn grain fermentation. This comparison was not robust to allocation, although the two corn stover processes were preferred over the corn grain process for most choices of allocation weights. More information is needed to decide conclusively which process is preferred. Extension of the framework to incorporate this additional information is proposed.

(564c) Hybrid Techno-Economic Modeling Tool for Greener Chemicals Supply Chains

Yao, Y., Northwestern University
Chang, Y., Northwestern University
Thwaites, F., Northwestern University
Masanet, E., Northwestern University
Originally presented on: 11/6/2013 16:05:00 - 16:30:00

The demand of chemical products is rapidly increasing due to the growth of global economy. At the same time, the chemicals industry is a major source of global greenhouse gas emissions (GHG) because of its high energy consumption(1). As such, reducing the energy and emissions footprints of chemicals is a critical priority for environmentalists, manufacturers and policy makers. Promoted by this question, green supply chain management is receiving increasing attention because it can help chemical industry to reduce the overall environmental intensity and thereby offer "greener" products. Most supply chain initiatives and related studies apply process-based life-cycle assessment (LCA) techniques to model and quantify the environmental impacts and sustainability performance of chemicals(2). These LCA studies help people to understand the current environmental footprints of many chemicals, but they are not able to answer the question: how and where can we reduce the life-cycle energy and pollution footprints of these chemicals? This question requires the support of techno-economic technology potentials analysis, which is a vibrant facilities engineering and analysis field that has operated largely independent of LCA to date(3, 4).

A comprehensive decision support tool incorporating techno-economic analysis methods and data into the life cycle assessment (LCA) framework is presented in this paper. The techno-economic analysis includes detailed characterizations of discrete technologies for reducing energy use and environmental impacts, their remaining installation potential in various plants and economic sectors, and their relative cost and environmental performance when deployed in combination. With this analysis, the model can identify and incentivize cost effective improvements for green supply chains within the chemicals industry to produce maximal energy reduction at minimal environmental burden. It can provide chemical manufacturers with comprehensive understanding of the greatest opportunities and potential cost requirements for supply-chain energy savings and environmental improvements. Also, the adoption of new process and material can be easily incorporated into this model, which can help designers to prescreen the sustainability performance of their design on a higher level. This tool also provides policy makers with cost-effective and efficient policy options to promote the development of more sustainable chemical product supply chains.

References

1. ICCA. 2009. *Innovations for Greenhouse Gas Reductions* ICCA

2. Goldberg A, Holdaway E, Reinaud J, O'Keeffe S. 2012. *Promoting Energy Savings and GHG Mitigation Through Industrial Supply Chain Initiatives* Institute for Industrial Productivity

3. Masanet E, Matthews HS, Carlson D, Horvath A. 2012. *Retail Climate Change Mitigation: Life-cycle Emission and Energy Efficiency Labels and Standards : Final Report*: University of California, Berkeley

4. McKinsey & Company. 2009. *Unlocking Energy Efficiency in the U.S. Economy.*, McKinsey Global Energy and Materials New York, New York

(564d) Design Decision-Making Under Uncertainty to Identify Sustainable Integrated Forest Biorefinery Strategies

Sanaei, S., Ecole Polytechnique de Montreal
Stuart, P. R., Ecole Polytechnique de Montreal
Originally presented on: 11/6/2013 16:30:00 - 16:55:00

Integrated biorefinery into existing pulp and paper facilities is a potential game-changing solution for the transformation of North American forestry industry, and the manufacture of an expanded product portfolio. However not all biorefinery strategies are sustainable and not have the same level of risk. Thus it is crucial that unsustainable strategies be systematically screened-out from the list of possibilities at the early design stage. The methodology presented in this study assesses candidate biorefinery alternatives from economic, competitiveness and environmental perspectives using different process systems analysis tools (e.g, techno-economics, Life Cycle Analysis, etc.) for both deterministic and uncertain conditions. The results of these assessments are used to evaluate a set of sustainability criteria, despite of the conventional analyses which generally use only short-term profitability metrics for decision-making. These conflicting criteria were aggregated into a unique sustainability score for each candidate biorefinery strategy by conducting a multi-criteria decision-making (MCDM) panel by which candidate strategies are ranked using a sustainability perspective.

However the candidate biorefinery strategies imply different levels of technology and market risk due to different sources of uncertainties which should be addressed in decision making. Thus to make a realistic strategic decision, the sustainability of biorefinery strategies should be assessed under uncertainty. In this study, different sources of uncertainty including market, technology and panel uncertainties are considered in biorefinery decision-making. In order to address this objective, a stochastic risk analysis method (Monte-Carlo Analysis) was used to develop probability distribution functions for sustainability criteria, which were then employed in a second MCDM panel. The proposed methodology in this study not only takes into account a complete set of sources of uncertainty, but also reflects the risk attitude of decision-makers in utility functions applying risk aversion theory. Comparing the results of deterministic and uncertain conditions shows that by evaluating uncertainty and quantifying risk attitude of panelists, the basis for the panel decision changes. This confirms the necessity of systematically addressing risk in biorefinery strategic decision making.

(564e) MILP-Based Clustering Method for Multi-Objective Optimization: Application to Environmental Problems

Oliva, D. G., Universitat Rovira i Virgili
Guillen-Gosalbez, G., Universitat Politecnica de Catalunya
Jimenez-Esteller, L., Universitat Rovira i Virgili
Originally presented on: 11/6/2013 16:55:00 - 17:20:00

Multi-objective optimization (MOO) has recently emerged as a useful technique in environmental engineering. One major limitation of this approach is that its computational burden grows rapidly with the number of environmental objectives, which causes difficulties regarding the computation and visualization of the Pareto solutions. In this work, we present several theoretical and algorithmic developments for grouping environmental objectives into clusters on the basis of which the multi-objective optimization can be performed, thereby facilitating the computation and analysis of the Pareto solutions. Particularly, we study how to group objectives of a MOO problem into meaningful clusters with the property that the minimization of any objective within a cluster will result in the minimization of the rest of objectives within the same cluster.

Our method is based on a novel mixed-integer linear program (MILP) that identifies in a systematic manner groups of objectives that behave similarly. The capabilities of the approach presented are tested through its application to several case studies, in which we compare the results of the MILP with those produced by a statistical method. The statistical method consists of a standard clustering algorithm.

The first case study deals with the design of petrochemical supply chains considering several environmental indicators simultaneously (Pozo et al., 2012). The second example deals with the design of bio-ethanol supply chains in Argentina considering 6 objectives (Kostin et al., 2012). The third example addresses the multi-objective optimization of hydrogen supply chains for vehicle use in Spain considering several Life Cycle Assessment impacts (Sabio et al., 2012). In the last example, we study an artificial dataset with 10 objectives and 41 points generated randomly. The goal is to test the performance of the statistical method when the objectives are poorly correlated.

Numerical results show that when the objectives are highly correlated, both methods (i.e., the MILP and the statistical method) lead to similar results, being the statistical approach the one with lower CPU times. In contrast, the statistical method leads to clusters with larger errors that differ from those produced by the rigorous MILP approach when the correlation between objectives is low. Our rigorous MILP-based approach is the only one that guarantees the minimum approximation error.

(564f) Emergy Method Analysis On Sludge Residues For Use From Waste-WATER Treatment Plant

Cano, N. A., Universidad Nacional de colombia
Gallego, D.,
Velasquez, H. I.,
Originally presented on: 11/6/2013 17:20:00 - 17:45:00

EMERGY METHOD ANALYSIS ON SLUDGE RESIDUES FOR USE FROM WASTE-WATER TREATMENT PLANT

ABSTRACT:

It was confirmed that the emergy analysis method allows comparison between the two alternatives for biosolids disposal while integrating disposition the energy, economic and environmental aspects. Therefore, by using the method of emergy analysis as a tool for sustainability assessment for environmental decision making, and understanding the sustainability of a disposal process that would reach a greater economic benefit with minimal environmental load, the actual benefit of biosolids "using biosolids as Soil Fertilizer" was evaluated and compared against a suitable alternative referenced in the literature as "energy recovery from Biosolids using fluidized bed gasification" under two scenarios: biosolids considered as a renewable resource and as an imported resource.

Being the best alternative according to the sustainability indices "Energy Recovery from Biosolids Using Fluidized Bed Gasification", ie, this alternative is energetically profitable, economically viable and environmentally acceptable; It does not mean that the actual use of the biosolids is economically or environmentally inappropriate. On the contrary, it does not produce environmental stress because its economic performance is not being taken advantage of at its full potential.

The analysis of emergetic levels not only allow us to evaluate the sustainability of a project, It also offers the possibility to improve the practice, so for example for the current use of biosolids the sustainability index allows us to say that a pretreatment of sewage sludge before its disposal improve to the alternative and become the most acceptable [1]

KEYWORDS: Emergy, sustainability, indicators, waste water, disposal, sludge

In Colombia, it is not uncommon to rely on plants for waste-water treatment purposes followed by the anaerobic stabilizing of biosolids as an alternative source of treatment. This practice is due to the many topographical limitations and conditions that do not allow mainstream treatment such as the use of oxidation ponds.

However, during waste-water treatment processing, large significative quantities of highly putrescible bio-solids with a higher energy concentration as well as organic charge/volume are produced; such is the case-example of the waste water treatment plant in San Fernando (an EEPPM-owned plant) where more than 70 tons of bio-solid material are generated, to be used to treat the 1.100m l/s of water needed to be released on a permanent basis [2].

Although bio-solid materials processed in this particular example is known for carrying similar traits to those found in urban areas, as well as being defined as neither toxic nor hazardous, they do in fact carry contaminants that required treatment and profitable use of such. Based on these findings, many alternative traditional methods for treatment and use of bio-solids have been implemented, those being for energy production, materials for construction purposes, restoration of natural habitat as well as agricultural uses [3]. However they have proven to be ineffective in their result.

In retrospect; it can be said that management of waste-water treatment is ultimately lacking and inconclusive, lest treatment of its additional waste product is carried-on in a sustainable fashion. defining sustainability of course, as the process of obtaining the highest financial return with a minimum effect in the environment; for such reason, the use of the emergy analysis method is a reliable gauging tool to measure level of sustainability to take full advantage of such, as proposed by [4] [5] [6] [7], as the most prominent and noteworthy method used in other countries [8] [9] [10] [11] [12] for sustainability based practices, in a manageable and equitative way in the economic, energy and environmental areas.

The emergy analysis approach is known as an analytical method, that through its integration of economy, ecology and thermodynamics concepts allows the assessment of economic, social and environmental impacts in a theoretical form, on development-driven projects, thus becoming a stronghold system in the fields ecological economy and environmental engineering.

As an emergy analysis method-based research, its ultimate intent is to weigh-in the efficient level of sustainability of any processing system used to make decisions concerning the environment. this method will be used in a way whereby a correct approach to evaluate all sludge residues found in any waste-water treatment plant can be identified in order to choose an assessment alternative that is ´applicable´. that is to say, an alternative that is ´´energy-suitable´´, ´´financially viable´´, and ´´environmentally friendly´´, in accordance to sustainability indicators; this research means to assess the use of the emergy analysis method as a sustainability appraisement tool of bio-solids production at a waste-water treatment plant in Colombia, while being able to satisfy all economic, environmental and energy levels of sustainability. Having as contrast, an already existing method of assessment found in the table of reference, allowing evidential proof of its utility on projects of a similar nature, as well as measuring if current alternative methods of waste-water treatment used in Colombia fit the standards of sustainability, by using the following parameters.

In order to meet such general objective, the application and results of the emergy analysis method used in a waste-water treatment plant in Colombia is analyzed, followed by an analysis of the system of sustainability. Then, by comparing the assessment of the two chosen alternatives of bio-solids used, through the estimate of emergy of both alternatives, the better method for bio-solid analysis is defined and accepted as the best one.

Thus, based on the sustainability indicators, the recovery of bio-solids energy through fluidized bed gasification becomes the best alternative for sustainability. That is to say, this alternative, meets all three standards of sustainability when compare to the alternative of using bio-solids as ground fertilizer.

REFERENCES

[1] Cano Londoño N. A. (2012). Análisis mediante el metodo emergetico de la disposición de los

lodos producidos en una planta de tratamiento de aguas residuales. (Aplicación a una PTAR en el Área Metropolitana del Valle de Aburrá). [Tesis MSc.]. Universidad Nacional de Colombia Sede Medellín.

[2] Velez, J. (2007). Biosolids: A problem or a solution? *Producción mas Limpia*, *2*, 57-71.

[3] Fytili, D., & Zabaniotou, A. (2008). Utilization of sewage sludge in EU: application of old and new methods—A review. *Renewable and Sustainable Energy Reviews*, *12*(1), 116-140.

[4] Agostinho, F., Diniz, G., Siche, R., & Ortega, E. (2008). The use of emergy assessment and the Geographical Information System in the diagnosis of small family farms in Brazil. *Ecological Modelling*, *210*(1-2), 37-57.

[5] Cherubini, F., Bargigli, S., & Ulgiati, S. (2009). Life Cycle Assessment (LCA) of waste management strategies: Landfilling, sorting plant and incineration. *Energy*, *34*(12), 2116-2123.

[6] Cohen, M. J., Brown, M. T., & Shepherd, K. D. (2006). Estimating the environmental costs of soil erosion at multiple scales in Kenya using emergy synthesis. *Agriculture, Ecosystems & Environment*, *114*(2-4), 249-269.

[7] Odum, H., & Odum, B. (2003). Concepts and methods of ecological engineering. *Ecological Engineering*, *20*(5), 339-361.

[8] Geber, U., & Bjo, J. (2001). The relationship between ecosystem services and purchased input in Swedish wastewater treatment systems — A case study. *Ecological Engineering, 18, 39-59.*

[9] Liu, Y., Kong, S., Li, Y., & Zeng, H. (2009). Novel technology for sewage sludge utilization: preparation of amino acids chelated trace elements (AACTE) fertilizer. *Journal of hazardous materials*, *171*(1-3), 1159-67.

[10] Siracusa, G., & La Rosa, A. D. (2006). Design of a constructed wetland for wastewater treatment in a Sicilian town and environmental evaluation using the emergy analysis. *Ecological Modelling*, *197*(3-4), 490-497.

[11] Zhou, J. B., Jiang, M. M., Chen, B., & Chen, G. Q. (2009). Emergy evaluations for constructed wetland and conventional wastewater treatments. *Communications in Nonlinear Science and Numerical Simulation*, *14*(4), 1781-1789.

[12] Zhou, S. Y., Zhang, B., & Cai, Z. F. (2010). Emergy analysis of a farm biogas project in China: A biophysical perspective of agricultural ecological engineering. *Communications in Nonlinear Science and Numerical Simulation*, *15*(5), 1408-1418.

(641a) Algal Biorefinery: Production of Biocrude Oil & Byproducts

Reddy, H. K., New Mexico State University
Selvaratnam, T., New Mexico State University
Muppaneni, T., New Mexico State University
Nirmalakhandan, N., New Mexico State University
Schaub, T., New Mexico State University
Dungen, B., New Mexico State university
Appuhamilage, N. S.,
Lammers, P., New Mexico State University
Voorhies, W., New Mexico State university
Deng, S., New Mexico State University
Originally presented on: 11/7/2013 8:30:00 - 8:52:00

We have developed a strategy needed for the production of sustainable algal biofuels with hydrothermal liquefaction of three different algal species. Hydrothermal liquefaction experiments with 10% solid loading of *Chlorella sp., Nannochloropsis sp.*, or *Galdieria sp.* were carried out between 180°C-300°C and 30 min. of reaction time to investigate the effects of HTL conditions on the important liquefaction reactions and final products. The liquid biocrude products were analyzed with GC-MS and FT-ICR for their chemical compositions, and micro bomb calorimeter for their high heating values. The composition of residual algal biomass was analyzed after each experimental run to identify the biochemical composition of the biomass. The residual water collected after liquefaction was also analyzed to quantify the nutrients and dissolved organic compounds in the water phase. This water phase produced in the HTL process was recycled for cultivation of *galdieria sp.* at different levels of dilution. An integrated processing strategy was developed for individual algal species based on the values of three products (bio-crudes, solid residues and water phase) obtained in the HTL process.

(641b) Biodiesel Production Through Microwave Assisted Transesterification of Microbial Cells

Cui, Y., Southern Illinois University

Originally presented on: 11/7/2013 8:52:00 - 9:14:00

One strain of oleaginous yeasts, Cryptococcus curvatus (ATCC 20509) has been studied to grow on several substrates including biodiesel production byproduct crude glycerol and sweet sorghum juice. After cultivation, yeast cells were heated under microwave radiation to extract lipid and produce biodiesel through in-situ transesterification.

Firstly, the yeast growth with crude glycerol was studied.On the second step, an optimization of yeast fermentation with crude glycerol was conducted. The resulting parameters of the response surface method optimization were used in a fed-batch fermentation where crude glycerol was automatically pumped in responding to exhausted oxygen levels in the fermentor.At the end of 12 days, the biomass density and lipid content were 44.53 g/L and 49%. Compared with our previous fed-batch experiment which was conducted under un-optimized condition, the yield of biomass and lipid increased 35.26% and 25.29%.

Considering its capability to utilize a broad range of substrates, C. curvatus was tested on its lipid productivity on sweet sorghum juice. When cultured in a fed batch process for 3 days, the final biomass density was 23.6 g/L with a lipid content of 51%. To extract lipids from cells in an effective and fast fashion, a domestic microwave oven was used with different solvents. With only methanol, a lipid yield of 33.2% of yeast cells was obtained in 4 min.

In the end, to convert yeast lipids to biodiesel directly without the step of lipid extraction,the in-situ transesterification method used microwave irradiation on the simultaneous extractionand transesterification of wet yeast biomass to biodiesel. Based on the experimental results and RSM analysis, the optimal conditions for this process were determined as: methanol to yeast biomass (v:w) ratio of around 50:1, catalyst concentration about 5 wt.%, and reaction time of 2 min. The biodiesel samples were analyzed with GC and the FAME content in biodiesel is about 50%. Compare the with other researches, this process proved to be a fast and easy method to produce biodiesel from wet yeast biomass.

(641c) Design Optimization By Response Surface Methodology for Continuous Fermentative Production of 1,3 Propanediol From Waste Glycerol By Product of Biodiesel Processes

Kanjilal, B., University of Connecticut
Noshadi, I., University of Connecticut
Intoci, N., University of Connecticut
Dowding, M.,
Hale, W.,
Brendel, B., University of Connecticut
Srivastava, R., University of Connecticut
Parnas, R., University of Connecticut
Originally presented on: 11/7/2013 9:14:00 - 9:36:00

The production of ~10 gallons of crude glycerol for every 100 gallons of biodiesel has led to a current glycerol excess of ~2.52Bn pounds per annum by the US biodiesel industry and non biodiesel sources. The bio-fermentation of crude glycerol to 1, 3-propanediol, an emerging intermediate platform in biorenewable chemicals with a nascent market of $575 million worldwide, as a value added product gives a recourse to capitalizing on the current over capacity of glycerol. This work presents the design optimization for a green and efficient process for the batch and continuous production of 1,3 propanediol from industrial waste glycerol using soil based bacterial inoculums. The optimal process parameters for the best possible yield of 1,3 propanediol were determined using a design of experiment by varying paramaters such as pH, glycerol concentration and yeast concentration. Response surface methodology (RSM), an important branch of experimental design, was used to find the relationship between experimental parameters and the yield and biomass growth in order to find the optimal conditions for the process. The optimization of 1,3 propandeiol production, in this way, with several inputs, helps to enhance the conversion in an economic manner. pH was seen to have the most profound effect on the process both in terms of yeild as well as bacterial growth. The optimised conditions were used to run a continous fermentation and arrive at a steady state for a given dilution rate, while the inhibitory solvent effects were mitigated by cycling between acetate and butyrate laced batches. While the addition of acetate enhanced biomass growth, butyrate addition enhanced 1,3 propanediol production with distinct inhibitory effects on the growth of bacteria. The effects were utilized to systematically maximize the survival of the mixed inoculum in a continuous run.

(641d) Review of Bio-Jet Fuel Conversion Pathways

Wang, W. C., National Renewable Energy Laboratory
Tao, L., National Renewable Energy Laboratory

Originally presented on: 11/7/2013 9:36:00 - 9:58:00

Bio-derived jet fuel has become a key subject in the aviation industry strategy to address the challenges from reducing operating cost, environmental performance and impacts to the global ecosystem. For producing a long-term renewable aviation fuel, economic and environmental sustainability are necessary. Researchers from oil refining industry, commercial aviation industry, government, biofuel companies, agriculture organizations and academic systems are looking forward to developing a commercial viable process that has not only low production cost but low GHG emissions. Additionally, the fuel produced from the process has to meet the ASTM specifications and is intended to be 100% drop-in for the replacement of current petroleum jet fuel. This study reviewed the current major technologies for converting bio-based materials into renewable aviation fuels. All the conversion pathways are classified to four categories: (1) syngas-to-jet, includes Fischer-Tropsch(F-T) synthesis and gas fermentation; (2) renewable-oil-to-jet, includes hydroprocessed renewable jet (HRJ) via plant oil and algal oil, catalytic hydrothermolysis (CH) and hydrotreated depolymerized cellulosic jet (HDCJ, also named pyrolysis) via pyrolysis oil; (3) alcohol-to-jet, includes the conversion of ethanol, butanol and methanol to jet fuel from lignocellulosic biomass feedstocks; (4) sugar-to-jet, includes fermentation of sugars to hydrocarbon or hydrocarbon precursors as well as catalytic conversion of sugars or sugar intermediate to fuels. Discussions for each pathway are primarily focused on the available feedstocks, conceptual process design, process economics, life-cycle assessments and commercial readiness, based on the literature studies. With the help of this comprehensive review on bio-derived jet fuel conversion technologies in recent literatures, it is believed that, whereas there are significant issues related to feedstock price and availability and energy consumption of the process, bio-derived jet fuels could progressively replace a great portion of conventional jet fuel required to meet the commercial and military demand.

(641e) Pellets From Lignocellulosic Materials: Briquetting Process Evaluation

Gonzalez, Y., Universidad Libre
Rincon, S., Universidad Libre
Camargo, G., Universidad Libre
Sierra, F., Universidad Nacional
Forero N, C. A., Universidad Nacional
Originally presented on: 11/7/2013 9:58:00 - 10:20:00

Colombia is a palm oil producer country and the implementation of biodiesel industry has generated huge quantities of the residues. On another hand, there are several regions in Colombia without a sufficient energy supply for a better and faster development. The purpose of this research is investigate the use of agro industrial residues coming from palm oil for obtain briquettes that can be used as fuel in remotes region in Colombia. Different parts of the agricultural waste they be employed in the manufacture of briquettes, the fibers and the seed. The biomass was characterized by ultimate analysis and in the densification process two different levels of pressure was employed. Finally densities, mechanical strength, heat of combustion were evaluated for the briquettes. The initial results shown that the waste from the seed materials presents a better performance for the densification process than the fiber materials and the briquettes elaborated a higher pressure exhibit more elevated calorific power

(641f) Pyrolytic Liquid Yields From Bamboo Species (Bambusa Oldhamii and Bamboo Angustifolia)

Pliego-Bravo, Y. S., Instituto Tecnologico de Orizaba
Cervantes- Pesado, L., INSTITUTO TECNOLOGICO DE ORIZABA
Bolanos-Reynoso, E., Instituto Tecnologico de Orizaba
Gonzalez-Garcia, P.,
Originally presented on: 11/7/2013 10:20:00 - 10:42:00

PYROLYTIC LIQUID YIELDS FROM BAMBOO SPECIES (*Bambusa Oldhamii and Bamboo Angustifolia*)

Abstract

A characterization of the species of bamboo (*Bambusa Oldhamii and Guadua Angustifolia*) with respect to compositional analysis, elemental analysis and thermochemical calorific using as a carrier gas, nitrogen, to identify desirable traits for subsequent use in a thermal conversion process such as pyrolysis, this energetic process provides greater value to convert biomass into a liquid fuel. Moisture content lignin degradation temperatures and ash content are indicators to determine the ideal raw material for conversion to a renewable process.

Introduction

Among the new alternatives that have been implemented for reducing the amount of emissions of CO, SOx, NOx, there being a source of biomass energy that partially replaces fossil fuels and decreases the concentration of these gases emitted into the atmosphere. (Zhang *et al*, 2007). Immediate analysis, elemental and calorific thermochemical analysis are useful in the understanding of the properties of the species to consider a material as an energy source, allowing to know the effective thermochemical process for producing a biofuel (Bridgwater, 2001 and Brown 2011). In this paper we characterize the species of bamboo (GA and BO) to propose an alternative to using craft or promote the efficient use of waste generated by the industry based on the characterizations mentioned above in order to obtain a liquid fuel

Methodology

The proximate analysis, elemental, calorific power and thermogravimetric were performed these are described below.

Materials

In lignocellulosic analysis were used: Ethanol analytical reagent grade (99.5% purity), analytical reagent grade sulfuric acid (98.2% purity), analytical reagent grade toluene, analytical reagent grade nitric acid (69% purity) all of J. T. Baker and distilled water. In thermochemical analysis was used 99% pure nitrogen and air.

Immediate analysis
Establishes the relative amounts of moisture (ASTM D 2867-99), volatile matter (ASTM D 5832-98), fixed carbon, ash (ASTM D 2866-99). Get the amount of carbon fixed subtracting the amounts of ash and volatile matter of a total of 100%.

Elemental analysis and calorific power
The amounts of C, H, N and S of the constituents of bamboo species were estimated experimentally using elemental analyzer (Vario EL III) based on moisture and ash free (LHC), the oxygen obtained by difference. The calorific value was estimated by the equation Reed and Das (1998) expressed in MJ / Kg.

Thermochemical analysis
Thermogravimetric analysis were performed with a non-isothermal model held in the TG Mettler Toledo 851e with software version Star. Samples between 6-10 mg were analyzed, studies were carried out using nitrogen and air, and before each experiment was performed a nitrogen purge for 10 min to all samples with a flow of 100 mL / min.

Getting pyroligneous liquids
For testing proceeded with an experimental 3^2 where the factors were time (15, 30 and 60 min) and the temperature (350, 500 and 600 ° C) for each of the species studied. These were carried out on a laboratory scale reactor with stainless steel fixed bed capacitors and nitrogen supply 200 mL / min.

Results and discussion

Moistures were between 16-17%, whereby the raw material was suitable for a thermochemical process. The high ash contents cause slag deposits on equipment deteriorating and decreasing the life thereof. The species studied showed percentages of 4.92 of ash for BO and 3.33 for GA making it attractive for use in these applications. Low fixed carbon content was related to the age of the material, in young bamboo lignin was removed easily due to low contents of total polysaccharides and cellulose, whereas an older kind more lignin is harder removing (Garcia *et al*, 2009).

The calorific value was obtained by correlations of Reed and Das (1998). In thermochemical analysis TGA curves showed that the decomposition of the species begins about 300 ° C, the displacement of the curves to the right with increasing temperature. For Bambusa oldhamii degradation stabilized 550 ° C to 600 ° C heating speeds of 15 and 25 ° C / min, while at 5 ° C / min stabilized at 500 ° C. In the case of Guadua angustifolia behavior was observed for more stable speeds of 5 and 15 ° C / min stabilizing at approximately 480 ° C whereas at 25 ° C / min from around 540 ° C. For older BO liquid yields were achieved at 600 ° C with 46 ± 0.5% in the times of 30 and 60 min, for the species GA best performance was obtained at 600 ° C for a time of 60 min of 43 ± 2%.

Referencias

[1] Bridgwater, A., V. (2001). *Progress in Thermochemical Biomass Conversion*. Backwell Science. Volume 2 IEA bioenergy.

[2] Brown R. C., *Thermochemical Process of Biomass*. Wiley, U.S.A. (2011).

[3] García, L, Marquez, F, Aguiar, L, Arauso, J, Carballo, L, Orea, U Zanzi, R (2009). *Rendimiento de los productos de la descomposición tÃ©rmica de la madera Eucalyptus saligna Smith a diferentes alturas del fuste comercial.* Revista Chapingo. Serie ciencias forestales y del ambiente, Vol. 15, Núm. 2. Julio_diciembre. pp. 147-154

[4] Higman C. y van der Burgt M., *Gasification*. Segunda edición. Elsevier. U.S.A. (2008).

[5] Reed T. B., y Das A., *Handbook of biomass downdraft gasifier engine system*. Solar Energy Research Institute, U.S. department of energy. U.S.A. (1998).

[6] Zhang Qi, Chang Jie , Wang Tiejun , Xu Ying (2007). Review of biomass pyrolysis oil properties and upgrading research. Energy Conversion and Management 48 (2007) 87–92.

(641g) The Impact of Ethanol and Iso-Butanol Blends on Regulated Emissions, Air Toxics, and Particle Emissions from SIDI Vehicles

Karavalakis, G., UC Riverside
Short, D., UC Riverside
Vu, D., UC Riverside
Durbin, T., UC Riverside
Asa-Awuku, A., UC Riverside
Originally presented on: 11/7/2013 10:42:00 - 11:04:00

Spark-ignition gasoline direct injection (SIDI) vehicles have been the subject of several studies, with much of this work has been done outside the U.S., as this is where GDI vehicles have been more prominent. The penetration of SIDI vehicles in the U.S. market is rapidly growing with most domestic automotive manufacturers offering direct injection versions in their existing models. The SIDI technology has emerged from the need to improve the thermodynamic efficiency (and thus reduce fuel consumption) and power output of spark ignited engines. The improved fuel efficiency compared to the conventional port fuel injection (PFI) vehicles could potentially provide the manufacturers the means to meet the target set in the U.S. on the fleet-average carbon dioxide (CO_2) emissions of future vehicles. It is foreseen that this category of vehicles will dominate the gasoline market, eventually replacing the conventional and less efficient PFI vehicles. However, SIDI engines are known to emit more particulate matter (PM) emissions than homogeneous-charge PFI spark-ignited gasoline engines and diesel engines equipped with diesel particle filters (DPFs) in the exhaust system. In particular, it is likely that SIDI vehicles will be higher than 3 mg/mile CARB LEVIII PM Standard through 2021, whereas PFI vehicles generally emit below 1 mg/mile, and often below 0.5 mg/mile, providing the potential to more than double PM emissions from new light-duty vehicles. The PM characteristics of direct injected spark ignition engines in comparison to conventional PFI engines can be mainly attributed to the injection characteristics. There is a growing consensus amongst the health experts that particles in the ultrafine range (smaller than 100 nm) are potentially more toxic and have greatest adverse health effects on human health.

Limited studies have investigated the influence of alternative fuels on the exhaust emissions, especially particulate emissions, from SIDI vehicles. The fuels employed in these studies were mainly low and mid-level ethanol blends. Currently, ethanol is the most widely used renewable fuel for transportation in the U.S. and with the push to use increasingly higher levels of renewable fuels, there has been an accompanying push to further increase the ethanol level in gasoline. The EPA Renewable Fuel Standard version 2 (EPA-RFS2) and the California Low Carbon Fuel Standard (CA-LCFS) are driving the U.S. market for the development of biofuels. The EPA-RFS2 requires that 36 billion gallons of renewable fuel are available in the U.S. market by 2022, with ethanol expected to make up the majority of this requirement.

Another pathway that could be used to reach the congressionally mandated biofuel volumes is the use of bio-

AUTHOR INDEX

AUTHOR INDEX

AUTHOR INDEX